# 47 Advances in Biochemical Engineering Biotechnology

Managing Editor: A. Fiechter

# Bioseparation

Volume Editor: G. T. Tsao

With contributions by
G. Belfort, A. D. Diamond, C. E. Glatz,
T. Gu, C. A. Heath, J. T. Hsu, J. H. T. Luong,
A. L. Nguyen, M. O. Niederauer, Y.-H. Truei,
G.-J. Tsai, G. T. Tsao

With 87 Figures and 9 Tables

Springer-Verlag
Berlin Heidelberg GmbH

ISBN 978-3-662-14989-8    ISBN 978-3-540-47211-7 (eBook)
DOI 10.1007/978-3-540-47211-7

Springer-Verlag Berlin Heidelberg 1992
Originally published by Springer-Verlag Berlin Heidelberg New York in 1992
Softcover reprint of the hardcover 1st edition 1992

Library of Congress Catalog Card Number 72-152360

Typesetting: Macmillan India Ltd., Bangalore-25

02/3020 5 4 3 2 1 0 Printed on acid-free paper

# Table of Contents

# Large-Scale Gradient Elution Chromatography

Yung-Huoy Truei, Tingyue Gu, Gow-Jen Tsai and George T. Tsao
School of Chemical Engineering, and Laboratory of Renewable Resources
Engineering, A. A. Potter Engineering Center, Purdue University,
West Lafayette, IN, USA

The goal of this chapter is to provide practical strategies for large scale separations by gradient elution chromatography. A detailed model has been developed for gradient elution systems considering interference effect, longitudinal diffusion, film mass transfer, intraparticle diffusion, mixing mechanism of the mobile phases, Langmuir-type adsorption and desorption kinetics. This detailed model can be solved by an efficient and robust numerical procedure. Hence, the optimizaton strategy of gradient elution has been developed through the calculation using this detailed model. This detailed model can precisely predict the band position, profile and width at various gradient concentrations, gradient periods, flowrates, and column lengths, in fair agreement with the experimental results. As a result of optimization, an optimal column length may exist. All the input parameters in this model have been either experimentally measured or estimated through empirical correlations. An alternative instrument for large-scale production using gradient elution has been suggested compared with conventional gradient elution instrument. The tolerance of the gradient elution processes to the fluctuation of input parameters has also been discussed.

Advances in Biochemical Engineering
Biotechnology, Vol. 47
Managing Editor: A. Fiechter
© Springer-Verlag Berlin Heidelberg 1992

## List of Symbols

| Symbol | Description |
|--------|-------------|
| a | constant in Langmuir adsorption equation $(-)$ |
| A | mobile phase component which has stronger affinity with the stationary phase |
| b | constant in Langmuir adsorption equation $(M)^{-1}$ |
| B | mobile phase component which has weak affinity with the stationary phase |
| C | concentration (M) |
| $C_b$ | eluate concentration (M) |
| CHY-A | a-chymotrypsinogen A |
| $C_m$ | eluent concentration (M) |
| CYT-C | cytochrome C |
| d | molecular diameter (cm) |
| dp | pore diameter (cm) |
| D | Brownian diffusivity $(cm^2\ s^{-1})$ |
| $D_b$ | axial dispersion coefficient $(cm^2\ s^{-1})$ |
| $D_p$ | intraparticle diffusivity $(cm^2\ s^{-1})$ |
| F | flowrate $(ml\ s^{-1})$ |
| k | film mass transfer coefficient $(cm\ s^{-1})$ |
| $k'$ | capacity factor $(-)$ |
| L | column length (cm) |
| LYS | lysozyme |
| Mr | molecular weight $(-)$ |
| $Pe_L$ | Peclet number of axial dispersion, $vLD_b^{-1}\ (-)$ |
| Re | Reynolds number, $2R_p\varepsilon_b v\rho\eta^{-1}\ (-)$ |
| RIB-A | ribonuclease A |
| $R_p$ | particle radius (cm) |
| Sc | Schmidt number, $\eta r^{-1}D^{-1}\ (-)$ |
| t | time (s) |
| $t_R$ | retention time (s) |
| $t_W$ | band width (s) |
| v | interstitial velocity $(cm\ s^{-1})$ |
| V | liquid volume (ml) |
| $V_m$ | internal volume of the mixer (ml) |
| $V_s$ | specific volume $(ml\ g^{-1})$ |
| z | $ZL^{-1}\ (-)$ |
| Z | axial coordinate (cm) |
| $Z'$ | proportional coefficient $(-)$ |

## Greek Letters

| | |
|--------|-------------|
| $\alpha$ | constant coefficient $(-)$ |

| | |
|---|---|
| $\beta$ | constant coefficient ( $-$ ) |
| $\delta$ | standard deviation of the Gaussian band ( $-$ ) |
| $\varepsilon_b$ | bed void fraction ( $-$ ) |
| $\varepsilon_p$ | particle porosity ( $-$ ) |
| $\eta$ | viscosity of the mobile phase (g cm$^{-1}$ s$^{-1}$) |
| $\gamma$ | constant coefficient ( $-$ ) |
| $\lambda$ | $dd_p^{-1}$ ( $-$ ) |
| $\rho$ | density of the mobile phase (g ml$^{-1}$) |
| $\tau$ | $tvL^{-1}$ ( $-$ ) |
| $\tau_{imp}$ | dimensionless time duration of the sample injection ( $-$ ) |

Subscripts

| | |
|---|---|
| i | ith component |
| 0 | initial value |

# 1 Introduction

This chapter is not intended to be a conventional review of gradient elution chromatography. Instead, the goal of this chapter is to provide practical strategies for large-scale separations using this method. Comprehensive reviews have provided its fundamentals and applications [1–4] for analytical purposes. In response to the increasing need for high purity bioproducts, advances in analytical liquid chromatography are being exploited for bioseparations [5]. Many of these bioproducts are proteins or other macro-molecules. However, most current theories and application strategies in gradient elution chromatography were developed for analytical purposes of small compounds, and they might not be appropriate for large-scale separations of macro-molecules, which will be generally described in this section and in detail in the remaining sections of this chapter.

Analyses are usually handled with small sample sizes and with dilute sample concentrations in the linear range of isotherms, with which the retention time and the band profiles of eluates are independent of the composition of the sample. By the same token, the elution bands in chemical analysis are usually treated as symmetrical Gaussian bands, whose band widths are always equal to $4\delta$, where $\delta$ is the standard deviation of the Gaussian band [4]. Under the assumption of Gaussian elution bands, it is a common belief that an increase in column length always improves separation performance. However, large-scale separations must be run with large sample sizes and/or with elevated sample concentrations, which have been shown to result in significant tailing of the bands with the concomitant loss of separation efficiency [6]. Thus, the nonlinearity of isotherms are often utilized in large-scale separations, in which the retention time and the band profiles of eluates, which are often asymmetrical, are dependent on the composition of the sample, which is called the interference effect [7]. For such asymmetrical elution bands of significant tailing, the common belief that an increase in column length always improves separation performance must be reexamined for large-scale separations. In an industrial scale operation, the greater length may mean an increased dispersion and thus affects the performance adversely. An effort to determine an "optimal" column length may be needed for large-scale separations.

The majority of current gradient elution theories emphasize the features regarding the chemical interaction between the stationary phase and the mobile phase [8–11]. Transport and kinetic problems in gradient elution systems are often overlooked, but can be significant in large-scale separations, especially for macro-molecules [12, 13]. Without considering the transport and kinetic effects, the band broadening and the band separation are difficult to elucidate [14]. Recently, the knowledge gained through studies in other fields of chemical engineering has been extended into the field of chromatographic separations. There is a large body of literature on band broadening due to the effects of transport and kinetics [15–22]. However, it is a challenge to develop a practical and realistic optimization strategy for large-scale separations by gradient elution chromatography considering the transport and the kinetic effects.

Consideration of the transport and the kinetic mechanisms makes the mathematical modeling very complex. Analytical solutions are usually unavailable for such a complex model [23]. As a result most scale-up processes of gradient elution chromatography have been carried out empirically [24]. The plate theory [25–27] and the lumped method [28, 29] have long been used to simplify the mathematical model. On the other hand, the simplifications which do allow analytical solutions often fail to reflect the reality of the system. For instance, the plate theory is limited to symmetrical Gaussian bands, and the lumped method is incapable of predicting the dynamic dependence of the chromatographic behavior on the input parameters, such as the flowrate, the particle size and the column length. Therefore, a detailed mathematical model considering the interference effect, the transport and the kinetic mechanisms must be used in predicting optimization of large-scale gradient elution chromatography. Recently, an efficient and robust numerical procedure has been developed for the solution of the complex mathematical model [30]. In addition, band broadening phenomena may be caused by different mechanisms including transport, kinetics, thermodynamics and in-column reactions, and these are often difficult to distinguish from one another [31, 32]. In other words, a detailed model with many adjustable parameters is usually able to fit most of the band profiles. Hence, the controlling mechanism must be determined before the detailed model is used. Otherwise, any further extrapolation and conclusions drawn from such a complex model without validating the controlling mechanism may be unrealistic.

The existing gradient elution instrumentation and procedures were also developed for analytic purposes. The simple extension of analytical instrumentation and procedures may not be sufficient for large-scale separations. For instance, when two or more mobile phases are mixed in gradient elution chromatography, air bubbles are often formed and then captured in the closed mixer, which may lead to distortion of gradient shape [33]. In the laboratory, various methods, including heating, helium and nitrogen gas purging, decompression, ultrasonification and using special degassing devices, are employed for removal of air from the mobile phases. These degassing methods are impractical in large-scale separations. An alternative instrument of gradient elution chromatography must be developed for industrial separations to prevent problems with air bubbles. Furthermore, the proportioning of mobile phases in gradient elution chromatography must be precisely controlled, otherwise the gradient shape may be distorted [1, 2, 4]. However, a variety of other causes can also lead to the distortion of the gradient shape [1, 2, 4]. These causes include the inaccurate flowrate of the pump, poor mixing of the mobile phases, and large hold-up volume of the mixer, as well as a large volume between the mixer and the column inlet. As a consequence, there is no question that highly precise and accurate gradient shapes are difficult to reproduce, particularly on various gradient devices [4, 34]. The distortion of gradient shape can be more serious in large-scale separations because industrial operations are usually not easily controlled as precisely as laboratory analyses. Therefore, the distortion of gradient shape must be solved in large-scale separations.

The retention relationship of the eluate concentrations and the eluent concentration describes how the eluent affects the retention of the eluates following the continual increase of the elution strength throughout the gradient period. Many conventional retention relationships developed for small molecules, such as the mass action law for small ions in ion exchange chromatography [35], have been extended to proteins. However, recent studies show the adsorption mechanism of proteins in ion exchange chromatography is not solely ion exchange [36–38]. One example is the significant hydrophobic interaction of macro-molecules in ion exchange chromatography, which has not been an important consideration for small compounds [39, 40]. Hence, Regnier has called the stoichiometric model as a non-mechanistic model and used the term electrochemical interaction chromatography (EIC) instead of ion exchange chromatography (IEC) for the adsorption of proteins in ion exchange systems [36]. Several empirical retention relationships of proteins have been developed [36, 39, 41].

The chromatographic procedures can be more precisely controlled in the laboratory than in an industrial setting. Therefore, the consistency of the gradient shape may not be easily achieved in industry. Other input parameters of chromatographic separations, such as feed concentrations, eluent concentration and pH value, can also vary from batch to batch in industrial operations. The tolerance of separation processes to the fluctuation of input parameters must also be considered in large-scale separations of gradient elution chromatography.

## 2  General Description

### 2.1  Overview

Gradient elution chromatography is a powerful tool for chemical analysis due to its broad range of retentivity, high peak capacity and short operation cycle [42]. The advantages of gradient elution chromatography are achieved by increasing elution strength during the gradient period, in contrast to the unchanged elution strength in isocratic elution chromatography. The continual increase of elution strength throughout the gradient period, known as a solvent gradient, is usually achieved by the proportioning of multiple mobile phases with a gradient former. Temperature gradient, flowrate gradient and column-material gradient or column switch (also called tandem columns) are alternatives to solvent gradient, but will not be discussed in this chapter. In solvent gradient, the gradient former programs the composition change of the mobile-phase mixture. Either the pumps or the valves, which must be programmable, are controlled by the gradient former in order to proportion the mobile phases (detailed in Sect. 3). The commonly used binary gradients are formed by two mobile phases, a weak component, called mobile-phase A in this chapter, and a strong one, called mobile-phase B. However, ternary or more complex gradients are also used particularly with the aim of eliminating the demixing effect of the mobile phases, which is caused by the incompatibility of the mobile phases [1, 4]. A mixer is also needed to mix the mobile phases and can be either a dynamic mixer or a static mixer (detailed in Sect. 3). Furthermore, the change of the mobile-phase composition change with time is called gradient shape. Gradient shape can be simply classified as continuous gradient and stepwise gradient, shown in Fig. 1. The continuous gradient includes linear gradient, also known as linear solvent strength (LSS) gradient [4] (see Figs. 4a and 4b), and nonlinear gradient (see Fig. 5). The stepwise gradient is composed of multiple steps of isocratic elution. Displacement chromatography, which uses a step-up of the displacer solution to displace the pre-loaded sample compounds, can be classified as a stepwise gradient chromatography. However, a complex gradient, such as multi-stepwise linear gradient (also known as segmented linear gradient) [43] (see Fig. 6), can be composed of the various simple gradients as well as isocratic steps. Usually, the eluent concentration increases during the continuous gradient period; while it decreases in hydrophobic interaction chromatography [40] (see Fig. 4b). In this chapter, only the linear gradient and the stepwise gradient will be discussed and compared due to the inconvenient complexity of other gradient techniques. Before the gradient starts, the column is equilibrated with the starting mobile-phase. After the end of a previous gradient run, the column must be completely reequilibrated with its initial mobile-phase before the next injection, usually by switching to its initial mobile phase rather than by a reverse gradient [1]. Incomplete equilibrium with the initial mobile phase after the prior run will

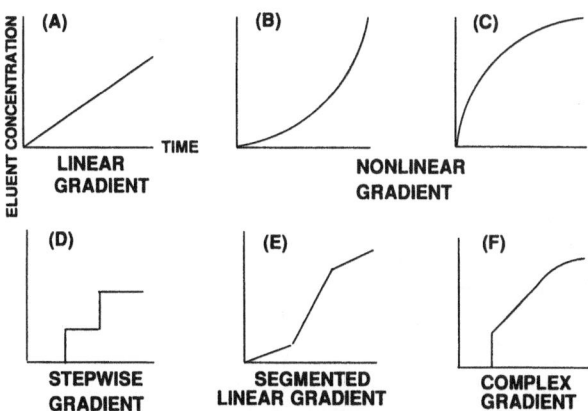

**Fig. 1.** Classification of gradient shapes

cause earlier elution and poor separation of the sample compounds in the next run. The sample compounds are usually dissolved in the initial mobile phase.

## 2.2 Advantages

In isocratic elution chromatography, the strongly retained sample compounds tend to tail and have late retention, shown in Fig. 2. To make these late-eluting bands sharper and elute faster with stronger elution strength, the weakly retained eluates might be poorly separated, shown in Fig. 3. However, in

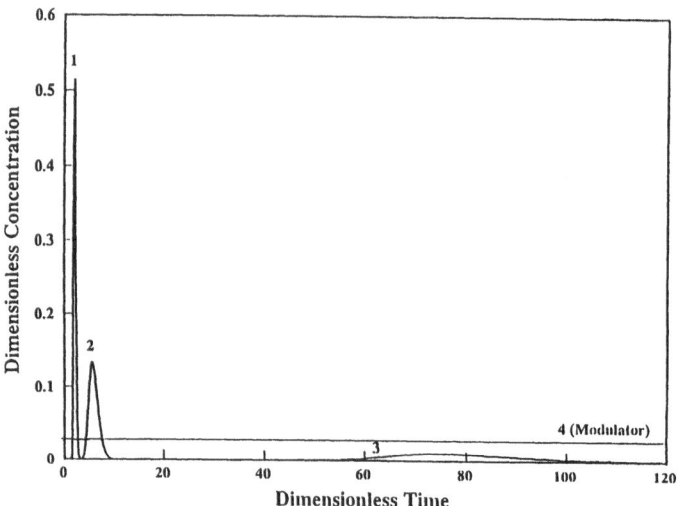

**Fig. 2.** Isocratic elution chromatogram calculated through the detailed model in the hydrophilic range of the retention relationship

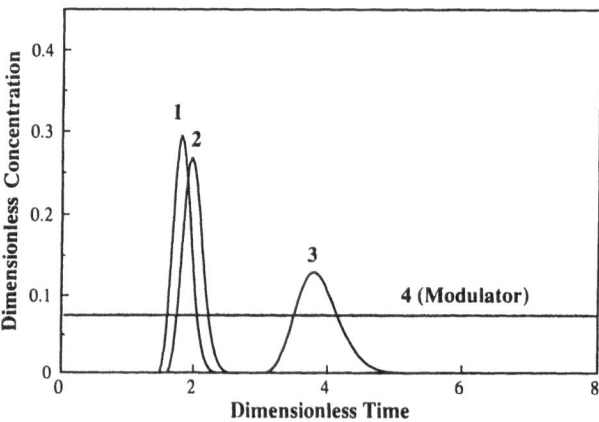

**Fig. 3.** Isocratic elution chromatogram at higher eluent concentration than in Fig. 2 calculated through the detailed model in the hydrophilic range of the retention relationship

gradient elution chromatography, the strongly retained eluates can be effectively stripped from the column by the continual increase of elution strength throughout the gradient, after the weakly retained eluates are well separated, as shown in Fig. 4. For this reason, the resulting bands are sharp, which means large peak capacity, and the separation cycle is short. Thus, gradient elution has great advantages versus isocratic elution in separating sample compounds which differ widely in retention on a chromatographic column, which is very common

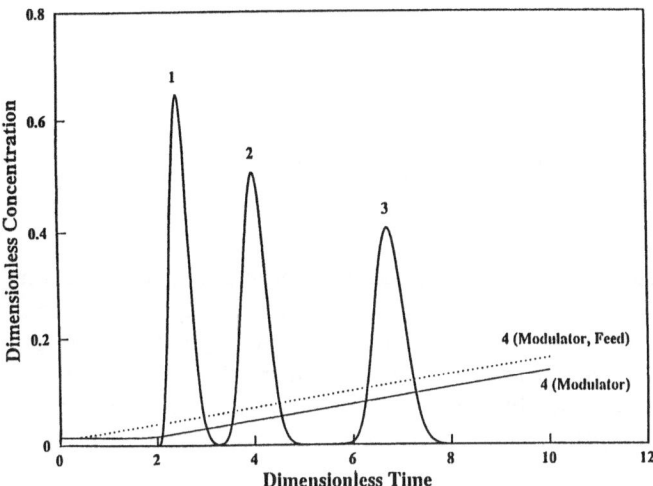

**Fig. 4a.** Linear gradient elution chromatogram calculated through the detailed model in the hydrophilic range of the retention relationship

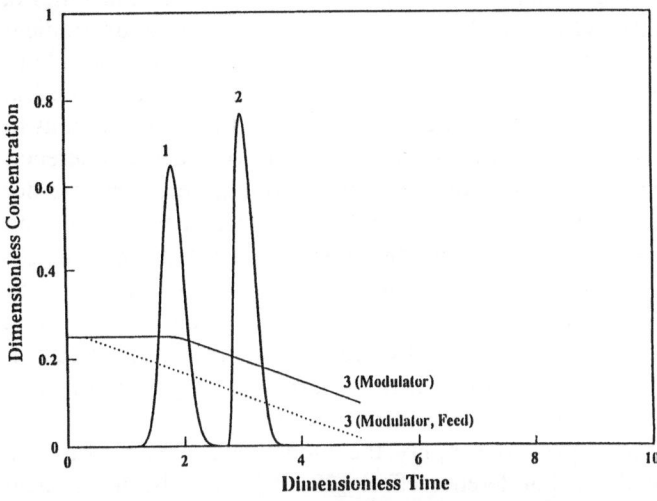

**Fig. 4b.** Linear gradient elution chromatogram calculated through the detailed model in the hydrophobic range of the retention relationship

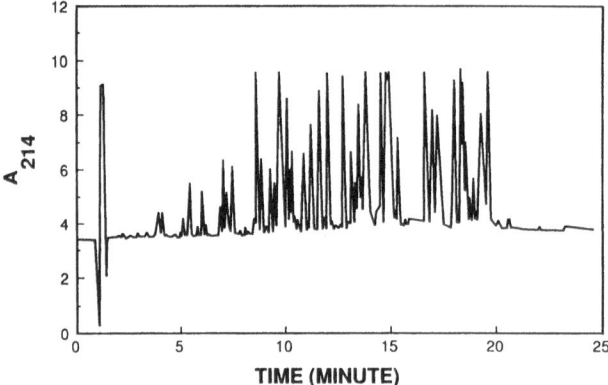

**Fig. 5.** Tryptic map of r-tissue plasminogen activator. 25 min linear gradient from 0 to 100% mobile phase (B); mobile phase (A): water/TFA, pH = 3, (B): acetonitrile: water = 60:40/TFA, pH = 3; column: Supelco RPC, LC18DB; detection: UV at 214 nm

in separations. Moreover, the gradient devices have been automated and commercially available. As a result, gradient elution chromatography has become a popular analytical technique in the laboratory. A case in point is the tryptic mapping of r-tissue plasminogen activator using gradient elution chromatography. Numerous peaks can be obtained in a single tryptic mapping chromatogram, shown in Fig. 5. Consequently, gradient elution chromatography provides fast and highly resolved separations, which also implies high loading capacity.

## 2.3 Disadvantages

Gradient elution chromatography is not a simple technique. The difficulties of reproducing the results, optimizing the conditions as well as scaling-up gradient elution separations are well known [34]. These difficulties occur because of both the theoretical and practical limitation of gradient elution. The theoretical calculation of gradient elution was limited by lack of the analytical solutions to the detailed model of gradient elution systems, that consider interference, transport and kinetic effects. Hence, further simplification is necessary in order to derive analytical solutions. The major simplifications include the assumption of Gaussian elution bands, linear chromatographic behavior with small sample sizes and dilute sample concentrations, simple retention relationships, and neglect of the transport and kinetic effects for the comparison of the existing models on gradient elution chromatography. However, as mentioned in Sect. 1, these simplifications have not been validated for large-scale separations, particularly of proteins. Several workers considered the transport or kinetic effect, but used lumping techniques to simplify the model and obtain analytical solutions, and ignored the interference effect [28, 29]. Likewise in gradient elution, isocratic elution also has the same theoretical limitation. However, it is much more difficult to calculate effluent profiles for gradient elution than for isocratic elution due to the complication of line-dependent mobile phase

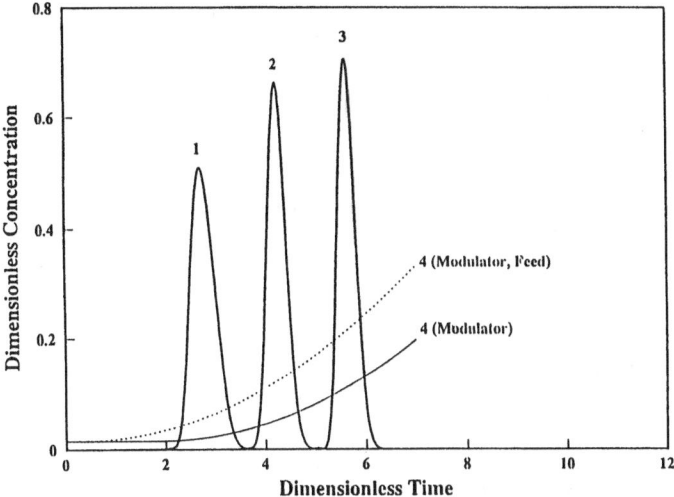

**Fig. 6.** Nonlinear gradient elution chromatogram calculated through the detailed model in the hydrophilic range of the retention relationship

composition. Numerical methods are currently the only solutions to the detailed model of the gradient elution system. However, an efficient and robust numerical procedure must be developed for such a detailed model, otherwise, the computational time will be expensive [30]. As a consequence, most scale-up processes of gradient elution chromatography have been carried out empirically [24].

The practical limitation of gradient elution chromatography is attributed to instrumental errors of the gradient devices. The basic requirements of gradient instrumentation is that they ensure consistency between the programmed gradient shape and the resulting gradient shape as it enters the inlet of the column. They require accurate and precise proportioning of mobile phases during the gradient run, and good mixing of the mobile phase mixture before it reaches the column [4]. However, in practice, a variety of causes for instrumental errors lead to distortion of the gradient shape. These include air bubbles, incomplete mixing of mobile phases, hold-up volume of the mixer and inaccurate flowrate of the pumps or valves over certain ranges of the gradient [44]. As previously mentioned, the air bubbles are often formed and captured by the closed mixer during the mixing of mobile phases [33]. Complete degassing of mobile phases by heating, helium and nitrogen gas purging, decompression, ultrasonification or the use of special degassing devices, is necessary to prevent air bubbles. The widely used reciprocating pumps need an additional pulse damper, and have limited accuracy in the 0–10% and 90–100% ranges of the gradient [4]. The larger the hold-up volume of the mixer, the more even is the mobile-phase mixture leaving the mixer [1, 4]. The hold-up volume between the mixer and the inlet of the column can also distort the gradient shape [4]. In addition, baseline shift or instability is another general problem in gradient elution, especially

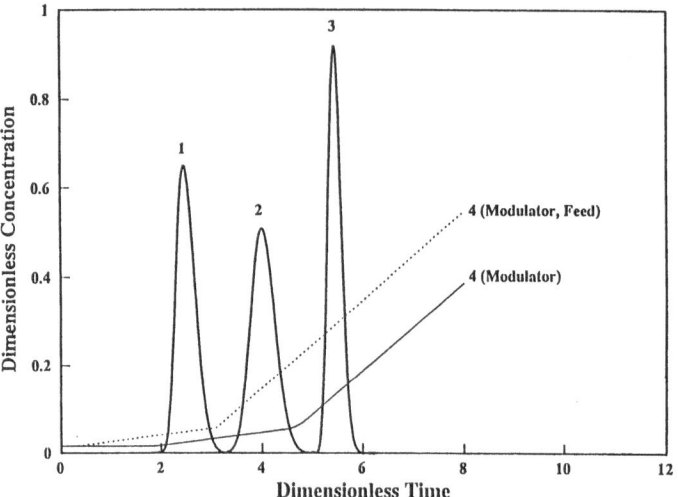

**Fig. 7.** Segmented linear gradient elution chromatogram calculated through the detailed model in the hydrophilic range of the retention relationship

when the mobile phases are incompatible. This problem will not distort the gradient shape, but will lead to the difficulty in quantifying the elution bands. The causes for baseline shift are complicated [45]. These instrumental errors result in difficulty in reproducing the gradient results, particularly on different gradient devices. These practical problems can be more serious in large-scale separations of gradient elution due to the rough conditions of industrial operations. The major goal of this chapter is to provide practical strategies to solve both the theoretical and the practical problems of gradient elution for large-scale separations. The results of gradient elution must be reproducible for repetitive industrial processes.

# 3 Equipment

## 3.1 Analytical Devices

The reproducibility of gradient elution results depends greatly on the performance of the instrumentation, as mentioned earlier. However, it is not easy to control precisely the composition of the mobile phase in gradient elution. In this context, several instrumental designs for gradient formation have been utilized [1, 4, 46–48]. Several workers have succeeded in reviewing and comparing the gradient devices [1, 49–51]. Most gradient devices have been commercially available and automated for laboratory analysis. These devices can be simply classified according to whether the mixing of mobile phases occurs at high pressure (see Fig. 8 for the major device of this type) or at low pressure (see Fig. 9 for the major device of this type). For high pressure mixing, the mixer is located downstream of the pumps and must be mechanically strong enough to undergo the high pressure generated by the pumps; while for low pressure mixing, the mixer is located upstream of the pump and consequently, mechanical strength requirements are less stringent. Furthermore, each mobile phase needs an individual pump for high pressure mixing, and only one pump is needed for low pressure mixing. For high pressure mixing, the proportioning of the mobile phases is carried out by controlling the flowrate of each pump, which must be programmable. Likewise, for low pressure mixing, programmable valves are used to perform the proportioning of the mobile phases. For both high and low pressure mixing, a controller, called the gradient former, is always needed to carry out the proportioning of mobile phases through pumps or valves. However, for a stepwise gradient, a single unprogrammable pump is sufficient, and the gradient former and the mixer are not necessary, although a flow-path switch is needed for changing the mobile phases. Four input parameters, which are gradient period, total flowrate, initial and final mobile-phase compositions, are usually fed into the gradient former for a linear gradient run. For a stepwise gradient run, the time and the mobile-phase composition of each step are the input parameters of the gradient former. Either a dynamic or a static mixer is also used for the mixing of the mobile phases. Both mixers are of the closed type. The dynamic mixer possesses active mechanical agitation, while

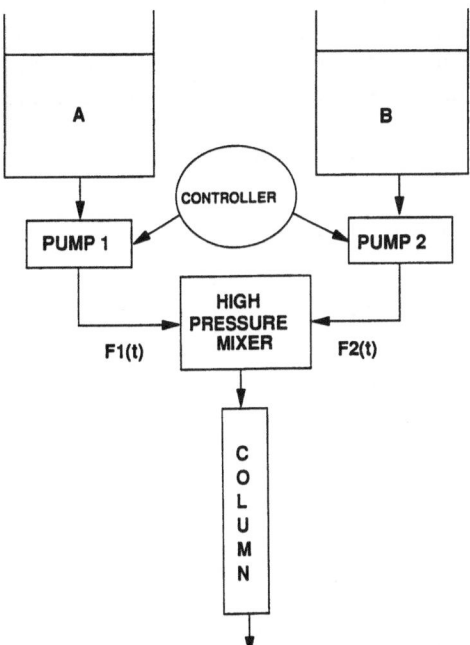

**Fig. 8.** High-pressure mixing gradient elution chromatographic instrumentation

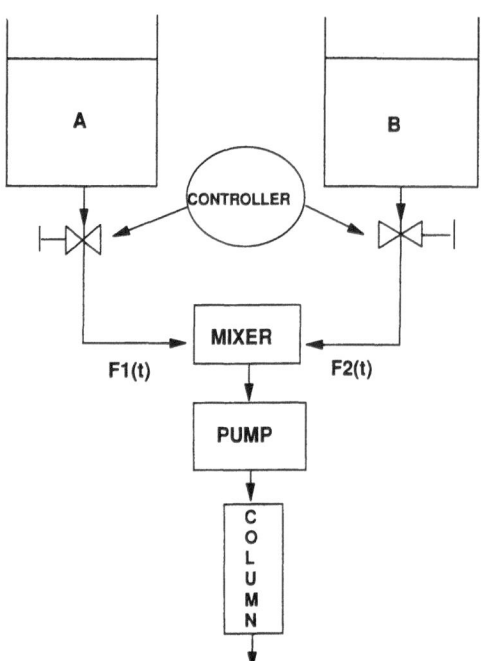

Fig. 9. Low-pressure mixing gradient elution chromatographic instrumentation

the static mixer does not. In principle, a T-connection can be used as a static mixer. However, a big hold-up volume of the static mixer is usually needed to dampen the turbulence arising from the mixing of the mobile phases. Figures 10 and 11 show how the turbulence can be built up and eventually distorts the programmed gradient shape due to the incompatibility of the mobile phases that results when a T-connection is used as a static mixer. In this case, the more

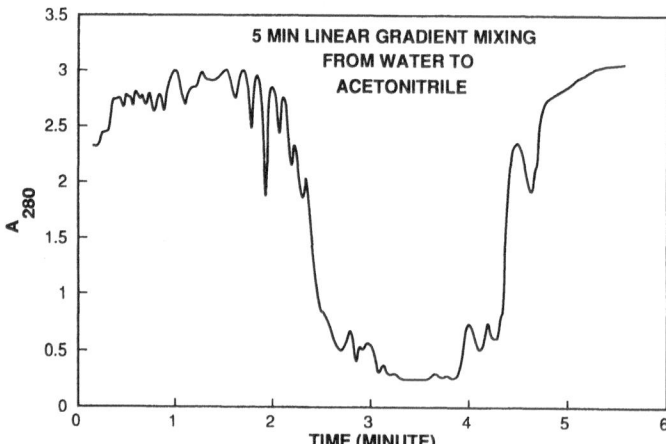

Fig. 10. Resulting gradient shape of 5 min linear gradient elution from water to acetonitrile with a T-connector as a static mixer at the flowrate of 1 ml min$^{-1}$

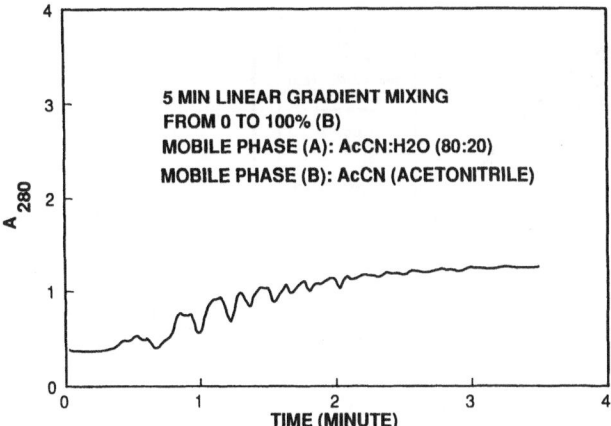

**Fig. 11.** Resulting gradient shape of 5 min linear gradient elution from a mixture of acetonitrile and water (80:20) to acetonitrile with a T-connector as a static mixer at the flowrate of 1 ml min$^{-1}$

incompatible the mobile phases are, the more turbulence can be generated. However, the hold-up volume of the mixer also can distort the programmed gradient shape [1, 4]. The larger the hold-up volume of the mixer, the more uniform is the mobile-phase mixture leaving the mixer [4], shown in Fig. 12. Figure 13 shows that for long gradient periods, the resulting gradient shape seems the same as the programmed except for a delay time, which is approximately equivalent to the average residence time of the mixer. But, for fast separations or short gradient period, the gradient shape is totally distorted by the hold-up volume of the mixer, as also shown in Fig. 13. The extent of the distortion of the gradient shape is proportional to the hold-up volume of the

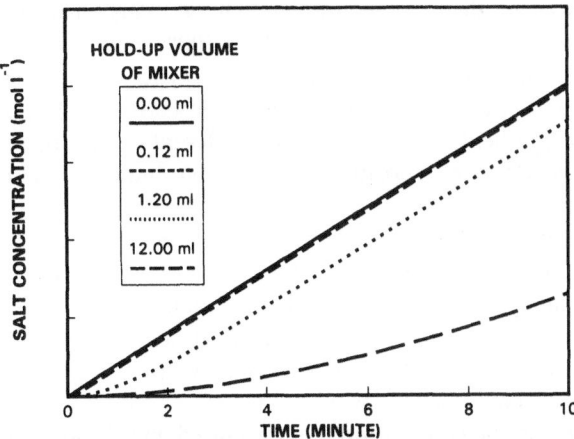

**Fig. 12.** Distortion of gradient shape by the hold-up volume of mixer in 10 min linear gradient elution from 0 to 2 mol l$^{-1}$ salt at the flowrate of 1 ml min$^{-1}$

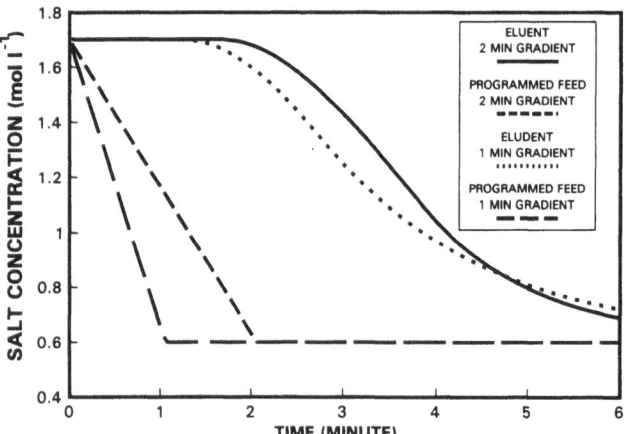

**Fig. 13.** Distortion of linear gradient shape by the hold-up volume, 1.2 ml, at various gradient periods from 1.7 to 0.6 mol l$^{-1}$ salt at the flowrate of 1 ml min$^{-1}$

mixer. Hence, the static mixer generally is sufficient only for very compatible mobile phases, and usually gives worst gradient results, even though it is cheaper than the dynamic mixer. As a consequence, a dynamic mixer which can provide perfect mixing with minimal hold-up volume will be the best choice for mixing if the cost consideration is not a problem. The hold-up volume between the mixer and the inlet of the column also can distort the gradient shape [4]. However, mixing mechanisms have been overlooked in the existing models of gradient elution chromatography. We believe that neglect of the mixing mechanism in the gradient system is one of the major reasons for the nonreproducibility of the gradient results and the difficulty in predicting and optimizing the gradient conditions [4, 34]. In addition to the mixing of the mobile phases, the inaccurate flowrate of the pumps also can distort the gradient shape. This problem especially occurs at low flowrate, i.e., in the 0–10% and 90–100% ranges of the gradient using the popular reciprocating pumps, as shown in Figs. 14 and 15 [1, 4]. In principle, Figs. 14 and 15 should be identical if the flowrate is accurate in the 0–10% range of the gradient. In fact, they are different. Thus, these gradient ranges must be avoided or a positive displacement pump should replace the widely used reciprocating pumps.

Apparently, the gradient device of high pressure mixing is more expensive than that of low pressure mixing due to the high-pressure mixer and additional pumps. However, it will usually prove to be worth the additional expense. Mobile phases usually contain some dissolved air from the atmosphere. When the mobile phases are mixed in the mixer, the resulting mixture are often supersaturated with dissolved air which is then released as bubbles. If air bubbles are released in the mixer, they are captured in the closed mixer and then pumped into the gradient system. Many problems including the distortion of the gradient shape arise from the formation of air bubbles. However, when the

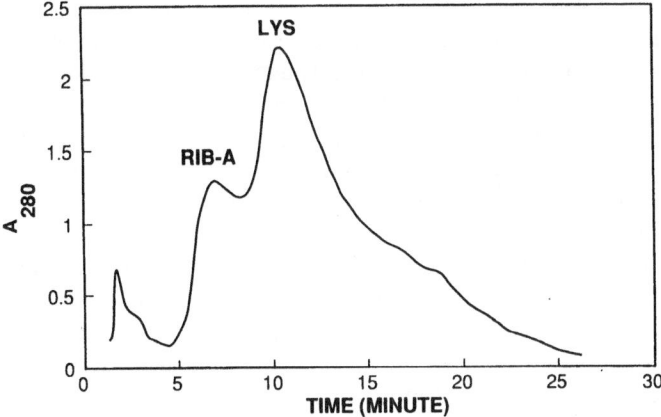

**Fig. 14.** Chromatogram of 15 min linear gradient elution from 0 to 100% mobile phase (B) on a column of Zorbax Bio-series WCX-300 (80 × 6.2 mm) at the flowrate of 1 ml min$^{-1}$; mobile phase (A): 10 mmol l$^{-1}$ ammonium sulfate in 20 mmol l$^{-1}$ phosphate buffer solution, pH 6, (B): 100 mmol l$^{-1}$ ammonium sulfate in 20 mmol l$^{-1}$ phosphate buffer solution, pH 6

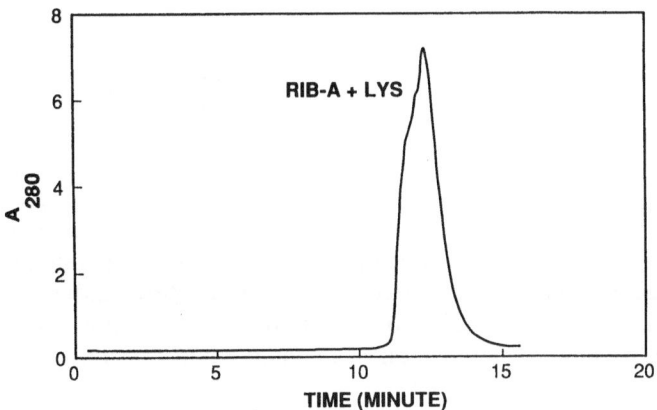

**Fig. 15.** Chromatogram of 15 min linear gradient elution from 0 to 10% mobile phase (B) on a column of Zorbax Bio-series WCX-300 (80 × 6.2 mm) at the flowrate of 1 ml min$^{-1}$; mobile phase (A): 10 mmol l$^{-1}$ ammonium sulfate in 20 mmol l$^{-1}$ phosphate buffer solution, pH 6, (B): 1 mol l$^{-1}$ ammonium sulfate in 20 mmol l$^{-1}$ phosphate buffer solution, pH 6

mixing of the mobile phases occurs under high pressure, the solubility of the resulting mixture is higher, and fewer air bubbles might be released. Even though the high pressure mixing cannot completely solve the problems with air bubbles, this problem is usually more severe for low pressure mixing, where extra effort in degassing is normally needed. This implies that the routine costs of gradient runs could be high regardless of the cheap initial instrument cost. The problem with air bubbles is particularly serious in reversed-phase chromatography because the mobile phases generally dissolve air to a widely different extent. Air bubbles can be formed not only in the mixer but also everywhere down stream of it. According to Bernoulli's equation [52], when the mobile phase passes from a wide cross-section through a narrow section, the increase in the velocity of the mobile phase will result in the decrease of the static pressure and the solubility of air. Then, air bubbles would be released due to the decrease of the air solubility. Therefore, the connector, especially when it is located in between the mixer and the inlet of the column, must have an internal cross-section area as uniform as possible. Complete degassing of mobile phases by heating, helium and nitrogen gas purging, decompression, ultrasonification or using special degassing devices, is necessary. Helium is widely used for degassing due to its low solubility in most liquids compared with air, and its use is treated as a routine operation cost. However, extensive degassing may vaporize the volatile mobile-phase components and change the mobile-phase composition. For instance, in reversed-phase chromatography, the composition of organic solvent and trifluoracetic acid in an aqueous solution can be lower than expected after extensive degassing.

## 3.2 Large-Scale Separation Devices

Apparently, the existing gradient instrumentation for analytical purposes still has many problems with instrument error. The major problem is that the resulting gradient shape departs from that programmed. This problem can be more serious in large-scale separations due to the more controlled conditions required in industrial operations, if the analytical instrumentation and procedures are simply extended to large-scale separations. Moreover, the conventional ways of degassing in laboratory analysis are impractical in industrial operations. In industry, the gradient shape must be consistent for repetitive industrial separation processes, and the formation of air bubbles must be prevented. Therefore, an alternative design of gradient instrumentation must be developed for industrial operations.

An alternative gradient system reported by Scott [53], shown in Fig. 16, has great advantages in industrial separations, but in contrast has some disadvantages in chemical analysis. This gradient system was reported before the currently strong interest in preparative chromatography, and has not been widely adopted. The Bio-Rad Model 385 gradient former used this idea of instrumentation except the use of gels for gradient formation [54]. This gradient system for high-performance columns is similar to a widely used gradient system for low-performance columns, shown in Fig. 17.

For linear gradient, $F_B = F_A \times 0.5$ (see Fig. 16), and

$$C_A = C_{A0} + (C_B - C_{A0})F_A t(2V_{A0})^{-1} \tag{1}$$

where C denotes the concentration, F the flowrate, t time, V the liquid volume in the vessel, and subscripts A, B and 0 denote vessel A, vessel B and the initial value, respectively. From this equation, it follows that the initial and the final

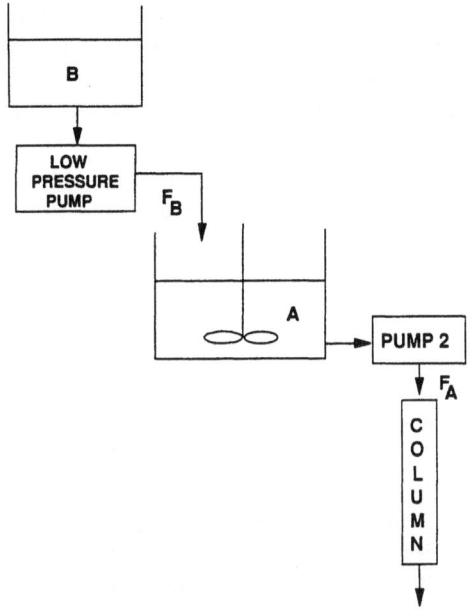

Fig. 16. Gradient elution chromatographic instrumentation of Scott [53]

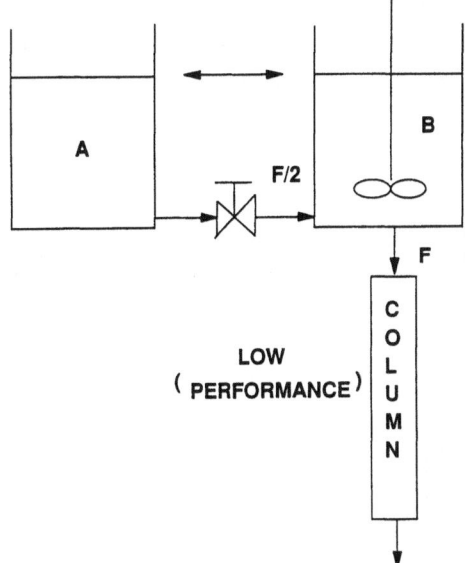

Fig. 17. Low-performance gradient elution chromatographic instrumentation

concentrations of gradient are $C_{A0}$ and $C_B$, respectively, and the gradient period is $(2V_{A0}F_A^{-1})$.

Its advantages in large-scale separations are:

– Mixing of mobile phases is carried out in an open vessel, which does not capture air bubbles. Hence, extensive degassing is not necessary.

- The mixing of mobile phases occurs in a vessel which originally contains mobile phase A. An additional mixer, especially one with high mechanical strength, is not needed.
- If the mixing in vessel A is efficient, the gradient shape will not be distorted by the hold-up volume of vessel A.
- The flowrates from two vessels remain unchanged. The proportioning controller and programmable valves are not needed. The pumps do not need to be programmable. There will be no problems with inaccurate flowrate in the 0–10% and 90–100% ranges of the gradient.
- The pump between two vessels does not need to be a high pressure pump. A multi-channel pump instead of two pumps can be used for this gradient system, shown in Fig. 18.

Since $F_B = F_A \times 0.5$ for a linear gradient, vessel A needs to be refilled after every gradient run. It is inconvenient for chemical analysis but does not pose a problem for standard industrial procedure. The gradient period needs to be calculated from $V_{A0}$ and $F_A$, which is impractical for analysis but is also acceptable for repetitive industrial processes. It has been found experimentally that this gradient system, compared to the conventional gradient systems does not exhibit the distortion of gradient shape due to inaccurate flowrate of the pumps, or incomplete mixing of the mobile phases and hold-up volume of the mixer [55], as shown in Fig. 19. The consistency between the resulting gradient shape and the programmed shape through this gradient system has also been experimentally demonstrated to be good [55] and is shown in Fig. 19. In

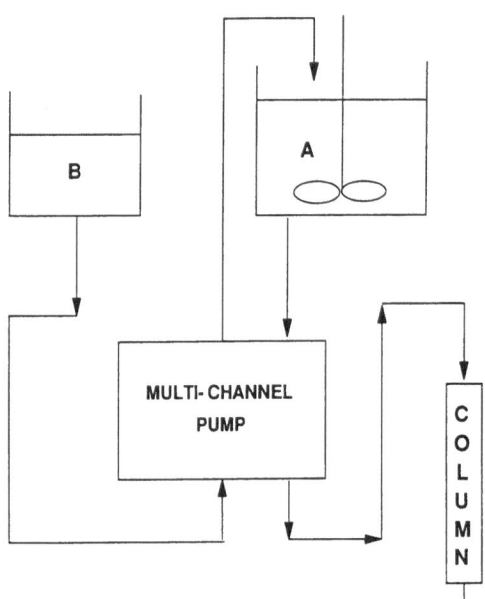

**Fig. 18.** Gradient elution chromatographic instrumentation of Scott [53] with a multi-channel pump

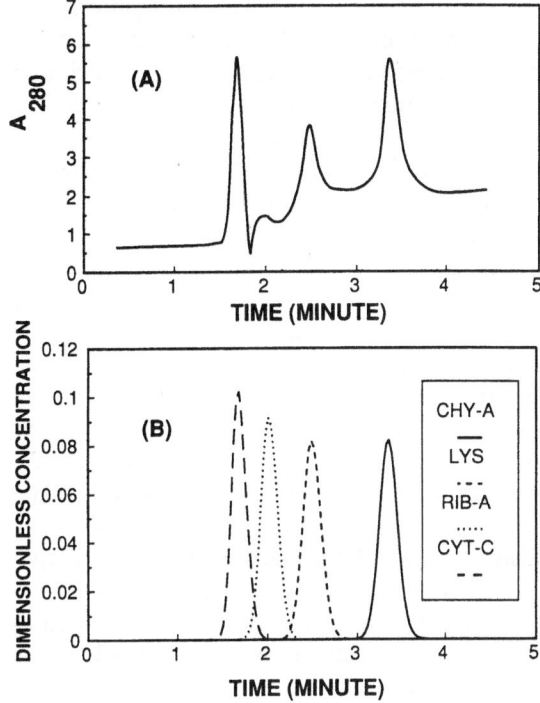

**Fig. 19.** Chromatograms of 2 min gradient elution from 1.7 to 0.6 mol $l^{-1}$ ammonium sulfate using the gradient elution instrumentation of Scott [53]. (**A**) experimental, (**B**) theoretical through the detailed model; column length: 8 cm; $v = 0.138$ cm s$^{-1}$

particular, the problem of air bubbles does not arise [55]. Consequently, this gradient system does not pose the practical problems with the instrumental errors which the conventional gradient systems have, is cost effective, and gives reproducible and consistent results. On the other hand, this gradient system is inconvenient for laboratory analysis.

# 4 Key Mechanisms

## 4.1 Retention Relationships

There are two major mechanisms which affect the retention of the eluates. The first one is the adsorption isotherm, which describes the relationship between the stationary concentration and the mobile phase concentrations. For multiple components, the multicomponent isotherm also describes the interference effect. The second one is the retention relationship of the eluate concentrations and the eluent concentration, which describes how the eluent affects the retention of the eluates.

Most existing models neglect the interference effect, which is insignificant for chemical analyses due to the small sample sizes and the dilute sample concentrations, but must be considered in preparative and large-scale gradient elution chromatography in which the column is often overloaded in terms of feed size and/or concentration [42]. Thus, the nonlinearity of isotherms is often utilized in large-scale chromatographic separations. This nonlinearity will cause the rear of the band broadened [32], while the leading edge of the bandit steeper. Thus, the band shape is often asymmetric and dependent on the composition of the mobile phase in large-scale chromatographic separations. However, a realistic adsorption equation can precisely elucidate the interference effect and the nonlinear response of the stationary phase on the band shape. The Langmuir adsorption equation is the most widely used for chromatographic separations due to its simplicity and capability of fitting many experimental results of chromatographic adsorption [56]. The extended multicomponent Langmuir adsorption equation is also capable of describing the interference effect [42]. However, the adsorption isotherm must be experimentally measured to validate the adsorption equation.

There are two major mechanisms with which the eluent affects the retentivities of the eluates. The first is the direct competition of the eluent with the eluates directly for binding sites on the stationary phase; here, the eluent is treated as a simple competing reagent. A case in point is the mass action law [35]. The second mechanism is the lowering of the adsorption equilibrium constant of the eluates by the eluent, for example, the ionic strength and the pH value have been used as the elution strength in ion-exchange chromatography [41, 57]. For small molecules, the first mechanism may sufficiently elucidate the elution phenomena in ion exchange chromatography, and a linear relationship between $\log(k')$ ($k'$ denotes the capacity factor) and $\log(Cm)$ ($Cm$ denotes the eluent concentration) with a positive proportional coefficient $\beta$ [58–60]. In reversed phase chromatography, Snyder [4] found that another linear relationship between $\log(k')$ and $Cm$ would be better. However, such linear relationships are true for proteins only over relatively narrow eluent concentration ranges, and the value of $\beta$ can be negative [61]. Other types of relationship for $k'$ and $Cm$ have also been developed (see Table 1). Some experimental plots of $\log(k')$ vs. $Cm$ or $\log(Cm)$ for proteins in ion exchange and reversed phase chromatography have been found to be nonlinear and exhibit minima [36, 37, 62]. Obviously, the role of eluent on the elution of proteins is more complex than a simple competitor [36, 37, 62] (see Fig. 20). Instead, both mechanisms are involved in the retention phenomena of proteins. Hence, an empirical correlation of $k'$ and $Cm$ must be used for proteins in gradient elution chromatography. An empirical correlation of $k'$ and $Cm$ has been developed for both hydrophilic interaction and hydrophobic interaction of proteins in ion exchange chromatographic systems, as follows [39].

$$\log(k') = \alpha - \beta\log(Cm) + \gamma Cm,$$

where $\alpha$, $\beta$ and $\gamma$ are constant coefficients.

**Table 1.** Comparison of models for gradient elution chromatography

| Reference | Correlation* of $k'$ v. $C_m$ | Isotherm | Interaction** type | Gradient method | Axial dispersion | Film mass transfer | Intraparticle diffusion | Kinetic effect |
|---|---|---|---|---|---|---|---|---|
| Pitt (60) | (1) | Linear | (1) | Linear | Yes | No | No | Linear |
| Jandera & Churacek (34, 77) | (1) & (2) | Linear | (2) | Linear | No | No | No | No |
| Hearn & Grego (62) | (1) & (2) | Linear | (2) | Linear | No | No | No | No |
| Armstrong & Boehm (78) | (2) | Linear | (2) | Linear | No | No | No | No |
| Kennedy et al. (79) | (1) | Linear | (1) | Linear | No | No | No | No |
| Furusaki et al. (24) | (3) | Linear | (2) | Linear | Yes | No | Yes | No |
| Yamamoto et al. (41) | (3) | Linear | (2) | Linear | No | No | No | No |
| D'Agostino et al. (80) | (5) | Linear | (2) | Linear | No | No | No | No |
| Ghrist & Snyder (43) | (2) | Linear | (2) | Linear & Stepwise linear | No | No | No | No |
| Markowski & Golkiewicz (81) | (1) & (2) | Linear | (2) | Stepwise | No | No | No | No |
| Martin (82) | (2) | Linear | (2) | Linear | No | No | No | No |
| Merengo et al. (83) | (6) | Linear | (2) | Linear | No | No | No | No |
| Antia & Horvath (42) | (1) & (2) | Langmuir (Multicomponent) | (2) | Linear | Yes | No | No | No |
| Kang & McCoy (84) | (3) | Linear | (2) | Linear | Yes | No | No | No |
| This Work | (4) | Langmuir (Multicomponent) | (2) | Stepwise linear & Nonlinear | Yes | Yes | Yes | 2nd Order |

* (1) $\log k' = \alpha - \beta \log C_m$　　(2) $\log k' = \alpha - \beta \log C_m$　　(3) $\log (k' - \alpha) = -\beta C_m$

(4) $\log k' = \alpha - \beta \log C_m + \gamma C_m$　　(5) $\log k' = \alpha_0 + \alpha_1 C_m + \alpha_2 C_m^2 + \cdots$　　(6) $k' = \alpha_0 + \alpha_1 C_m + \alpha_2 C_m^2 + \cdots$

** (1) Modulator competes with eluates for binding sites, but does not affect the $k'$ value of the eluates

(2) Modulator affects the $k'$ values of the eluates, but has negligible adsorption on the stationary phase

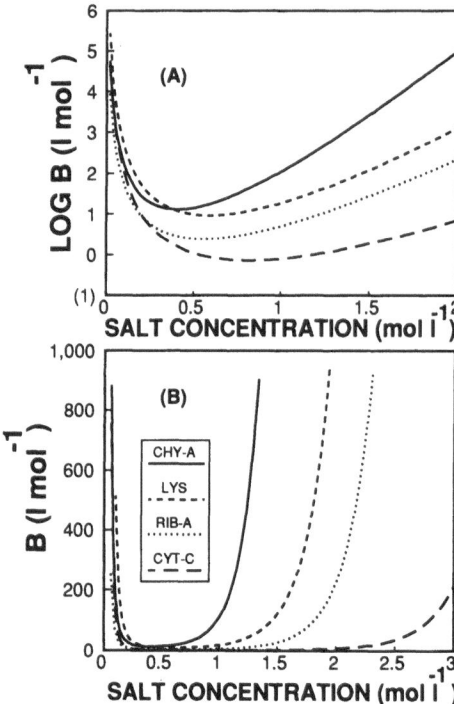

**Fig. 20.** Retention relationships between b and Cm at pH 6 on a column of Zorbax Bio-series WCX-300 (80 × 6.2 mm)

However, many conventional retention relationships developed for small molecules, such as the mass action law for small ions in ion exchange chromatography, have been inappropriately extended to proteins. First of all, the mass action law cannot account for the hydrophobic interaction of proteins in ion exchange chromatography. Second, the Langmuir adsorption equation is equivalent to the mass action law when the characteristic valence in mass action law is equal to one. But, the characteristic valences of proteins are usually not one [63]. Third, the characteristic valences of proteins vary during the process [63], however, they are assumed as constants in most models which employ the mass action law. Fourth, the mass action law cannot explain slow desorption due to the low possibility of simultaneous dissociation of all of the multiple binding sites of proteins [64].

Four proteins were chosen as the eluates in a recent study of gradient elution chromatography [55]: α-chymotrypsinogen A from bovine pancreas (CHY-A), lysozyme from chicken egg white (LYS), ribonuclease A from bovine pancreas (RIB-A) and cytochrome C from horse heart (CYT-C). Ammonium sulfate was chosen as the eluent. A cation exchange system was chosen for these proteins due to their high pI values. The retention relationships of these proteins and ammonium sulfate [55] were plotted in Fig. 20, which fit the empirical correlation developed by Melander and Horvath [39]. The multicomponent Langmuir adsorption equation was also used in the study of gradient elution chromatography [55].

## 4.2 Mass Transport

Even though the mixing of mobile phases can distort the gradient shape [4], no current model in gradient elution chromatography considers the mixing mechanism. However, a dynamic mixer can be modeled as a CSTR with an internal volume Vm (ml) [55]. The programmed gradient shape entering the mixer and the resulting gradient shape from this mixer were plotted in Figs. 12 and 13. Figure 12 shows that the distortion of the gradient shape is increased with the hold-up volume of the mixer. For a long gradient period, the resulting gradient shape looks the same as the programmed gradient shape except for a delay time, which is approximately equivalent to the average residence time of the mixer, as shown in Fig. 13. However, for a fast separation or a short gradient period, the gradient shape is totally distorted by the hold-up volume of the mixer, as also shown in Fig. 13. Figure 21 illustrates the deviation of the predicted chromatogram without mixing from that with mixing. Hence, we believe that the neglect of the mixing mechanism in the gradient system is one of the major reasons for the nonreproducibility of the gradient results and the difficulty of predicting the result and optimizing the gradient conditions [55].

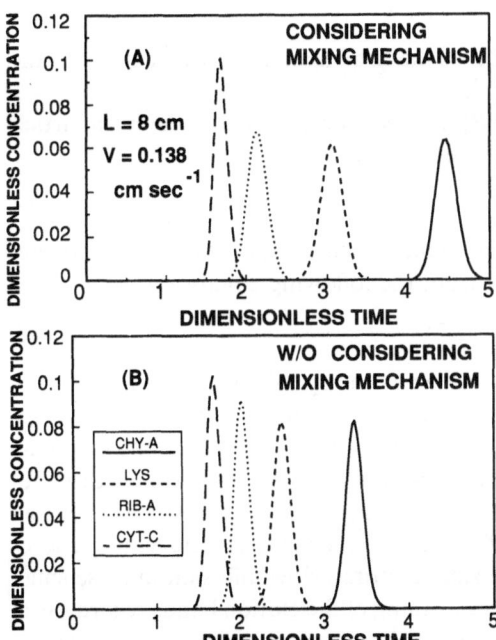

**Fig. 21.** Comparison of chromatograms of 2 min linear gradient elution from 1.7 to 0.6 mol l$^{-1}$ ammonium sulfate considering the mixing mechanism (A), and without considering the mixing mechanism (B); column length: 8 cm; v = 0.138 cm s$^{-1}$

Axial dispersion, film mass transfer and intraparticle diffusion are considered as the key mass transfer mechanisms. The eluates are carried by the convective flow of the mobile phase. Along with the convective flow, the injection band can be broadened by axial dispersion. Axial dispersion is caused by Brownian diffusion, eddy diffusion, the boundary layer effect, channeling (if the column was packed improperly), and the wall effect [27]. Then, the eluates need to penetrate through a boundary film on the outer surface of the particles. For most chromatographic particles, the eluates move from the entrance of the pores to the intraparticle surface solely by the intraparticle diffusion. However, for perfusable materials [65], convective flow also occurs within the pores. For macromolecules such as proteins, the hindered diffusion regarding the relative ratio of the molecular size of the eluates to the pore size must be considered for the intraparticle diffusion [66]. The mass transfer coefficient can be estimated by empirical correlations as follows.

The correlation of Chung and Wen [67] can be used to estimate $Pe_L$, the Peclet number of axial dispersion:

$$Pe_L = (0.2 + 0.11 Re^{0.48}) L / (2 R_p \varepsilon_b) \tag{2}$$

where the Reynolds number $Re = 2 R_p \varepsilon_b v \rho \eta^{-1}$, $\varepsilon_b$ is the bed void fraction, v is the interstitial velocity (cm s$^{-1}$), r is the density of the mobile phase (g ml$^{-1}$), $R_p$ is the particle radius (cm), and $\eta$ is the viscosity of the mobile phase (g cm$^{-1}$ s$^{-1}$).

The correlation of Wakao, et al. [68], can be used to estimate k, the film mass transfer coefficient, for the film mass transfer (cm s$^{-1}$):

$$2 R_p k D^{-1} = 2 + 1.45 Re^{0.5} Sc^{3^{-1}}, \quad Re < 100 \tag{3}$$

where $R_p$ is the particle radius (cm), D is the Brownian diffusivity (cm$^2$ s$^{-1}$), and the Schmidt number $Sc = \eta \rho^{-1} D^{-1}$.

The correlation of Yau et al. [66], is used to estimate $D_p$, the intraparticle diffusivity, for the intraparticle hindered diffusion:

$$D_p = D(1 - 2.104\lambda + 2.09\lambda^3 - 0.95\lambda^5) 2.1^{-1} \tag{4}$$

where $\lambda = d \times d_p^{-1}$, d is the molecular diameter (cm) and $d_p$ is the pore diameter (cm). The parameter d is calculated from the following equation [69]:

$$d = 2(MrVs \times 1.246 \times 10^{-23}) 3^{-1} \tag{5}$$

where Mr is the molecular weight, and Vs is the specific volume (ml g$^{-1}$).

Recent research [55] has indicated that the distribution of pore size versus intraparticle surface area is broad, as shown in Fig. 22. However, the manufacturers claim the pore size of their products to be narrow, based on the distribution of pore size versus pore volume, shown in Fig. 23. The pore size distribution is important when the hindered diffusion is significant. As long as the macromolecular eluates can penetrate the smaller pores, slow diffusion in these small pores must be a major cause of the broadening of the elution bands regardless of the existence of the larger pores. Therefore, the existence of the small pores must

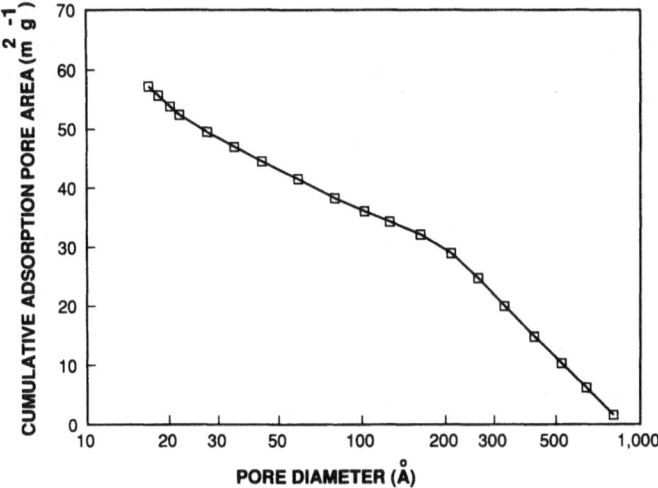

**Fig. 22.** Distribution of pore size vs. cumulative adsorption pore area of Zorbax Bio-series WCX-300

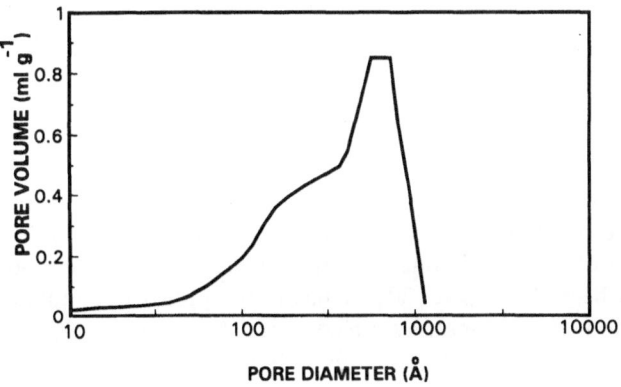

**Fig. 23.** Distribution of pore size vs. adsorption pore volume of Zorbax Bio-series WCX-300

be avoided for chromatographic materials (including perfusable materials, which contain macro-pores [65]). The pore volume distribution can be precisely measured by the method of nitrogen gas adsorption [70] for the estimation of the intraparticle porosity, $\varepsilon_p$.

## 4.3 Adsorption and Desorption Kinetics

The adsorption of the eluates is often fast compared with the mass transfer rate, but can be slow enough to broaden the injection band. A case in point is the affinity chromatography in which the eluates may need numerous collisions before they adsorb on the surface due to specific orientation requirements of the collision [71]. Slow desorption occurs more often than the slow adsorption [71]. Slow desorption can be caused by the multiple binding sites of proteins; it is not easy for the macromolecules to dissociate simultaneously at all binding sites during desorption [64]. Some cases of slow desorption have been discovered in high affinity chromatographic systems [13]. Slow adsorption and desorption can broaden elution bands and reduce separation performance. When the adsorption or the desorption of the eluates is slow, adsorption and desorption kinetics must also be studied in addition to the adsorption equilibrium, and the adsorption and the desorption rate constants also must be experimentally measured. There is no empirical correlation available for the measurement of adsorption and desorption rate constants. Slow adsorption or desorption can be examined easily using frontal technique with a mini- or micro-column at increasing flowrate [55]. When the flowrate is increasing, a minimal breakthrough time results, this is equivalent to the inclusion volume if slow kinetics is the rate limiting step, and is equivalent to the exclusion volume if the mass transfer is the rate limiting step. A mini- or micro-column is used in this experiment to allow for the high pressure drop expected at an elevated flowrate.

# 5 Optimization

A detailed mathematical model of gradient elution chromatography considering interference effect, longitudinal diffusion, film mass transfer, intraparticle diffusion, mixing mechanism of the mobile phases, Langmuir-type adsorption and desorption kinetics has been developed [30]. It has been applied to simulate large scale gradient elution chromatography. An empirical retention correlation of b and Cm, $\log(b) = \alpha - \beta\log(Cm) + \gamma Cm$, where b is the equilibrium constant in the Langmuir adsorption equation, for proteins in an ion-exchange system was used [39]. The hydrophobic interaction range of eluent concentration is chosen due to the higher relative affinities of the proteins in this range than in the hydrophilic interaction range (see Fig. 20). All the input parameters have been either experimentally measured or estimated through empirical correlations [55]. This model can predict band positions with a relative error of less than 5% at various initial and final eluent concentrations (see Figs. 24 and 25), flowrates (see Figs. 24, 26 and 27), gradient periods (see Figs. 24, and 28–30), and column lengths (see Fig. 31), in linear gradient elution chromatography [55]. Stepwise gradient elution chromatography has also been studied with various stepwise periods and stepwise eluent concentrations (see Figs. 32–36), and compared with linear gradient elution chromatography experimentally and theoretically using the detailed model [55]. However, the required long computation time could be the bottle-neck in using this detailed model. Hence,

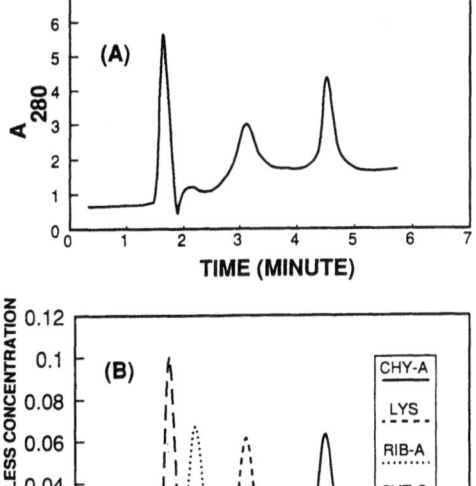

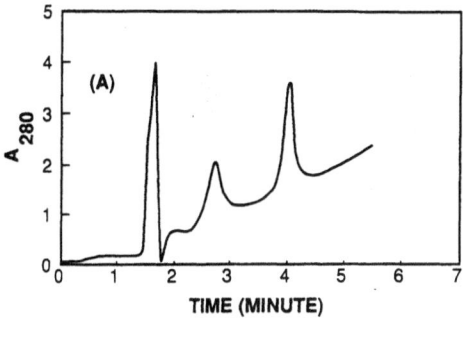

**Fig. 24.** Chromatograms of 2 min linear gradient elution from 1.7 to 0.6 mol l$^{-1}$ ammonium sulfate. (**A**) experimental, (**B**) theoretical through the detailed model; column length: 8 cm; v = 0.138 cm s$^{-1}$

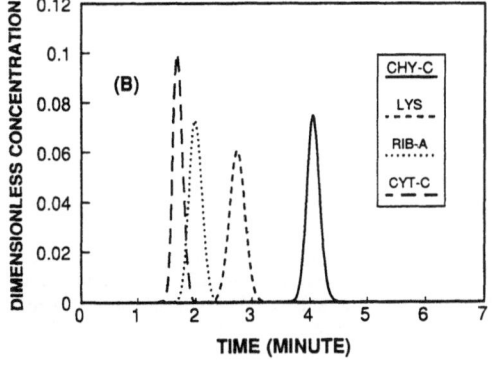

**Fig. 25.** Chromatograms of 2 min linear gradient elution from 1.6 to 0.4 mol l$^{-1}$ ammonium sulfate. (**A**) experimental, (**B**) theoretical through the detailed model; column length: 8 cm; v = 0.138 cm s$^{-1}$

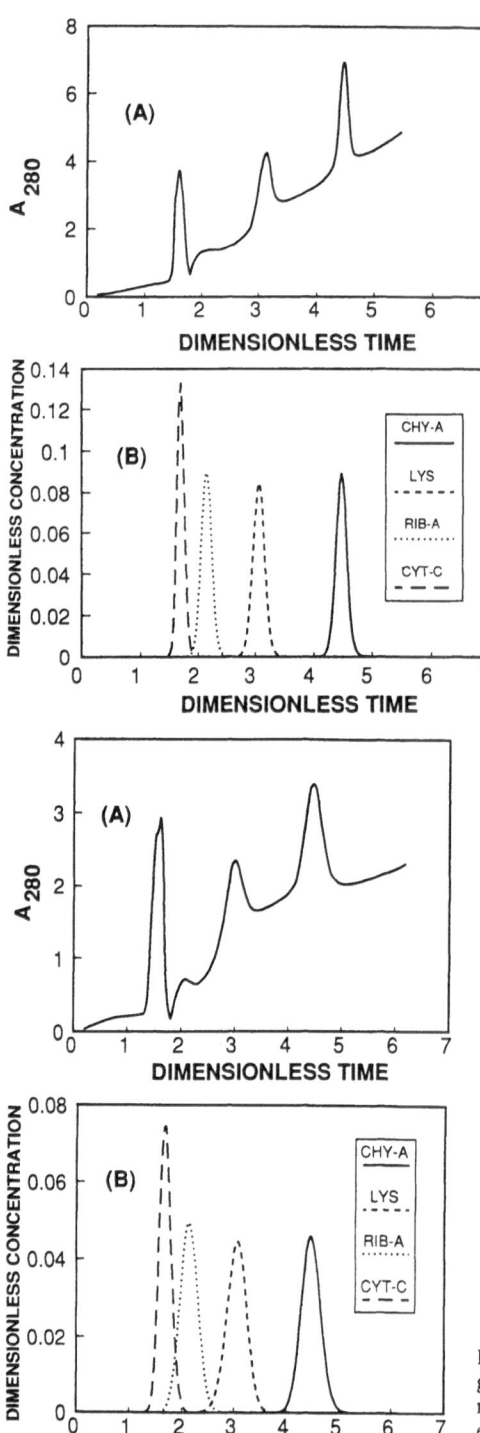

**Fig. 26.** Chromatograms of 4 min linear gradient elution from 1.7 to 0.6 mol l$^{-1}$ ammonium sulfate. **(A)** experimental, **(B)** theoretical through the detailed model; column length: 8 cm; v = 0.069 cm s$^{-1}$

**Fig. 27.** Chromatograms of 1 min linear gradient elution from 1.7 to 0.6 mol l$^{-1}$ ammonium sulfate. **(A)** experimental, **(B)** theoretical through the detailed model; column length: 8 cm; v = 0.276 cm s$^{-1}$

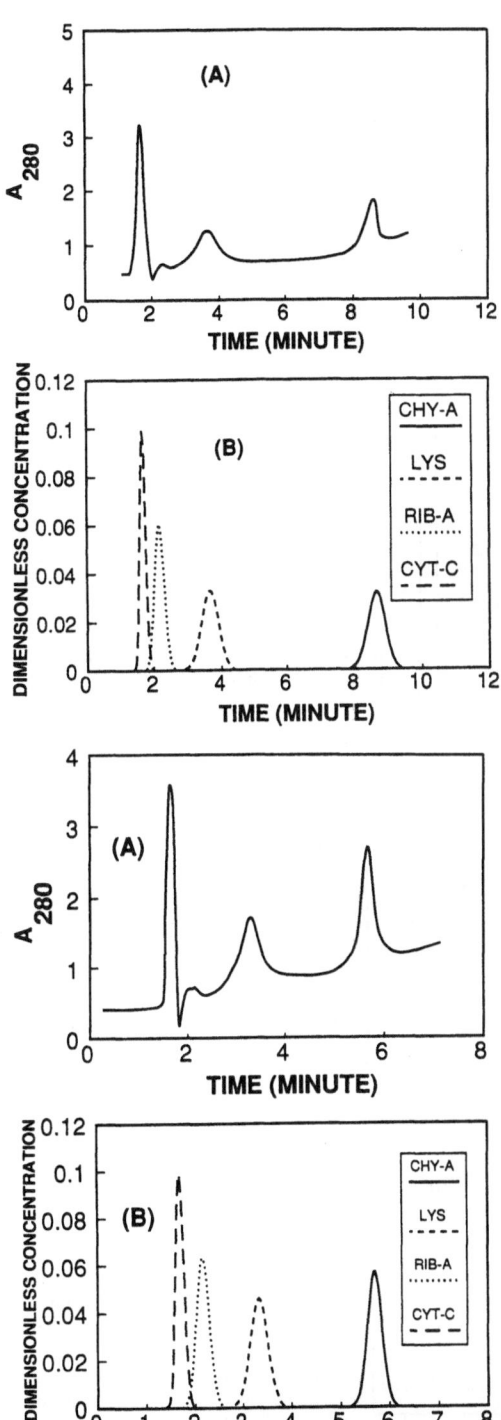

**Fig. 28.** Chromatograms of 10 min linear gradient elution from 1.7 to 0.6 mol l$^{-1}$ ammonium sulfate. (**A**) experimental, (**B**) theoretical through the detailed model; column length: 8 cm; v = 0.138 cm s$^{-1}$

**Fig. 29.** Chromatograms of 4 min linear gradient elution from 1.7 to 0.6 mol l$^{-1}$ ammonium sulfate. (**A**) experimental, (**B**) theoretical through the detailed model; column length: 8 cm; v = 0.138 cm s$^{-1}$

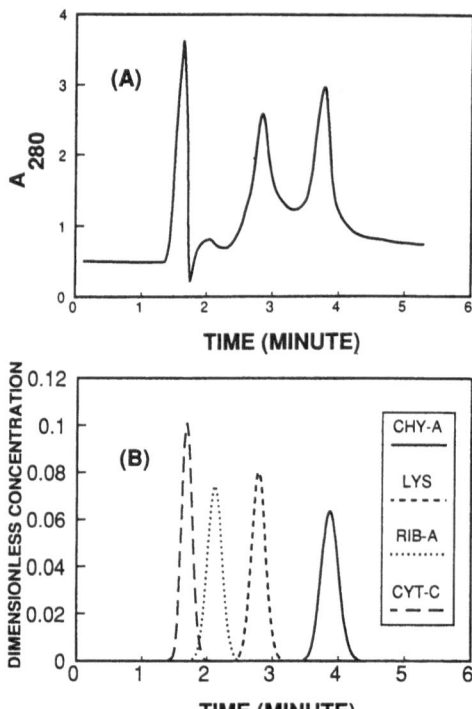

**Fig. 30.** Chromatograms of 1 min linear gradient elution from 1.7 to 0.6 mol l$^{-1}$ ammonium sulfate. (**A**) experimental, (**B**) theoretical through the detailed model; column length: 8 cm; v = 0.138 cm s$^{-1}$

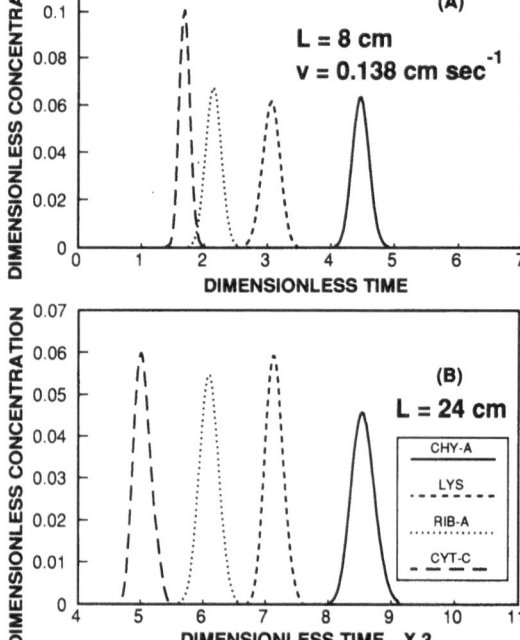

**Fig. 31.** Comparison of chromatograms of 2 min linear gradient elution from 1.7 to 0.6 mol l$^{-1}$ ammonium sulfate at various column lengths calculated through the detailed model. (**A**) 8 cm, (**B**) 24 cm; v = 0.138 cm s$^{-1}$

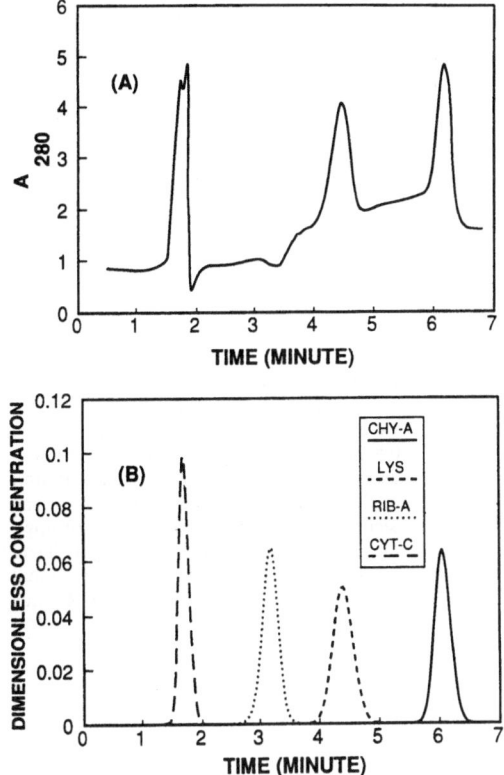

Fig. 32. Chromatograms of multi-stepwise gradient elution of 1 min 2 mol l$^{-1}$, 2 min 1.3 mol l$^{-1}$, then 0.6 mol l$^{-1}$ ammonium sulfate sequentially. (A) experimental, (B) theoretical through the detailed model; column length: 8 cm; v = 0.138 cm s$^{-1}$

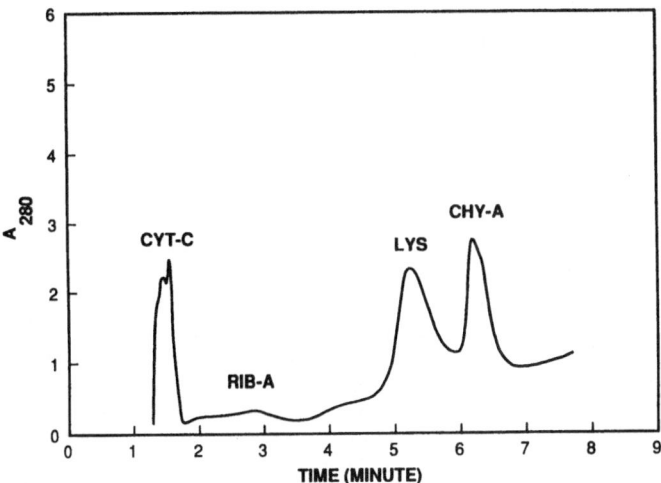

Fig. 33. Chromatogram of multi-stepwise gradient elution of 1.5 min 2 mol l$^{-1}$, 1.5 min 1.4 mol l$^{-1}$, then 0.6 mol l$^{-1}$ ammonium sulfate sequentially on a column of Zorbax Bio-series WCX-300 (80 × 6.2 mm) at pH 6 and the flowrate of 1 ml min$^{-1}$

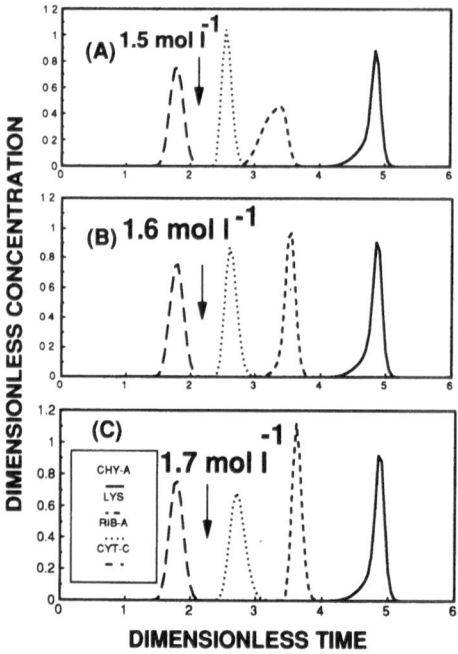

**Fig. 34.** Chromatograms of multi-stepwise gradient elution of 0.5 min 2 mol $l^{-1}$, 1.1 min 1.6 mol $l^{-1}$, 1.4 min 1.1 mol $l^{-1}$, then 0.5 mol $l^{-1}$ ammonium sulfate sequentially using the gradient elution instrumentation of Scott considering the fluctuation of the second-step eluent concentration through the detailed model; column length: 8 cm; v = 0.138 cm s$^{-1}$. (**A**) minus the maximal fluctuation, 0.1 mol $l^{-1}$, (**B**) normal, (**C**) plus the maximal fluctuation

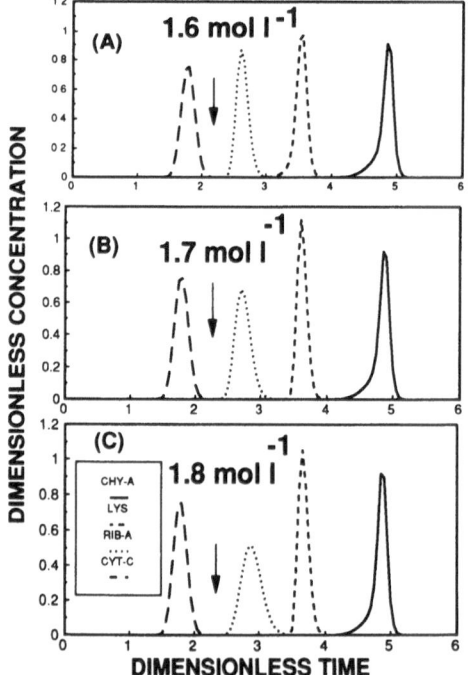

**Fig. 35.** Chromatograms of multi-stepwise gradient elution of 0.5 min 2 mol $l^{-1}$, 1.1 min 1.7 mol $l^{-1}$, 1.4 min 1.1 mol $l^{-1}$, then 0.5 mol $l^{-1}$ ammonium sulfate sequentially using the gradient elution instrumentation of Scott considering the fluctuation of the second-step eluent concentration through the detailed model; column length: 8 cm; v = 0.138 cm s$^{-1}$. (**A**) minus the maximal fluctuation, 0.1 mol $l^{-1}$, (**B**) normal, (**C**) plus the maximal fluctuation

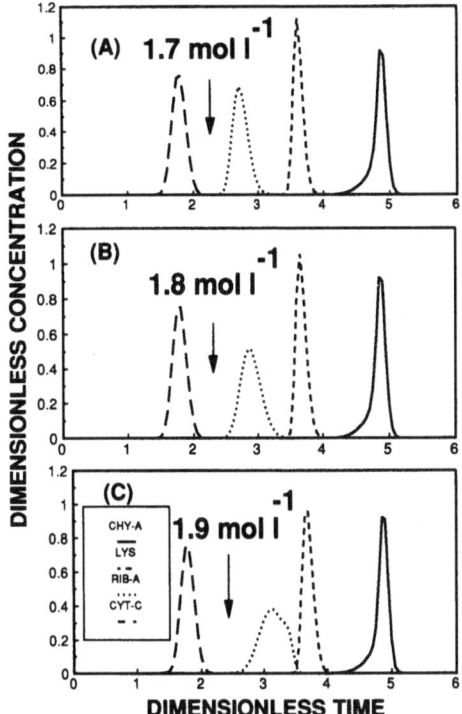

**Fig. 36.** Chromatograms of multi-stepwise gradient elution of 0.5 min 2 mol l$^{-1}$, 1.1 min 1.8 mol l$^{-1}$, 1.4 min 1.1 mol l$^{-1}$, then 0.5 mol l$^{-1}$ ammonium sulfate sequentially using the gradient elution instrumentation of Scott considering the fluctuation of the second-step eluent concentration through the detailed model; column length: 8 cm; $v = 0.138$ cm s$^{-1}$. (**A**) minus the maximal fluctuation, 0.1 mol l$^{-1}$, (**B**) normal, (**C**) plus the maximal fluctuation

a practical strategy for optimization has been developed using this detailed model, as illustrated in Fig. 37.

After the cation exchange column, the pH value of the mobile phase and the hydrophobic interaction range of the eluent concentration have been chosen, the final eluent concentration of the linear gradient can be determined from the eluent concentration at which the weakest eluate has the minimum b. Then the shortest acceptable gradient period (see Fig. 38) is chosen as the first guess to calculate the ideal retention time of eluates at various initial eluent concentrations of the linear gradient (see Fig. 39) through the concentration wave equation [32]:

$$(dz d\tau^{-1})_i = \{1 + [(1 - \varepsilon_b)\varepsilon_p\varepsilon_b^{-1}] + [(1 - \varepsilon_b)(1 - \varepsilon_p)f'(C_{bi})\varepsilon_b^{-1}]\}^{-1}$$

(6)

where $f'(C_{bi}) = d[a_i C_{bi}(1 + \sum b_j C_{bj})^{-1}]d(C_{bi})^{-1}$, $C_b$ is the eluate concentration, a is a constant in the Langmuir adsorption equation, z is the dimensionless axial coordinate, $\tau$ is the dimensionless time, $\varepsilon_b$ is the bed void fraction, and $\varepsilon_p$ is the particle porosity. Then, the ideal distances of adjacent eluate peaks can be obtained (see Fig. 40). A good separation needs the peak distances to be larger than the dilution ratio of the feed impulse by the pore liquid (see Fig. 41), as $(1 - \varepsilon_b)\varepsilon_p\varepsilon_b^{-1} + \tau_{imp}$.

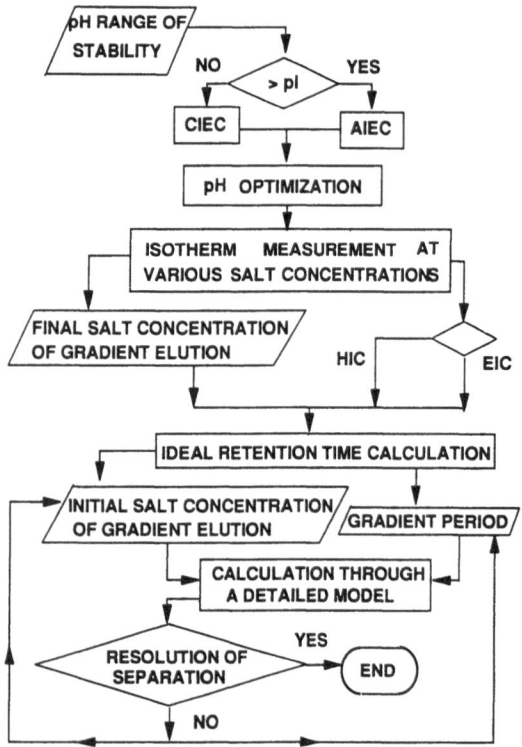

**Fig. 37.** Flow-sheet illustration of optimization strategy of linear gradient elution chromatography

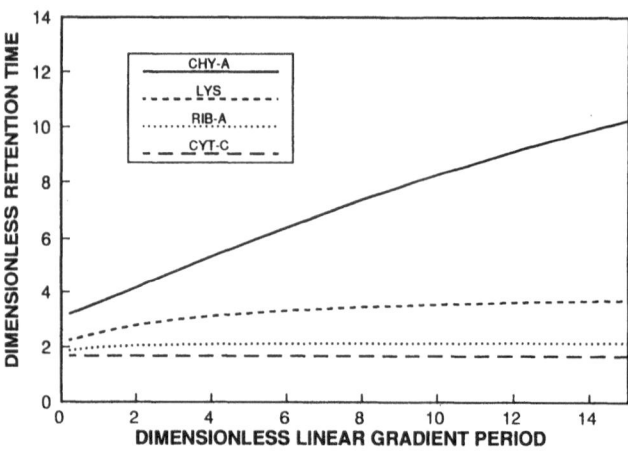

**Fig. 38.** Retention time vs gradient period of linear gradient elution chromatography calculated through the ideal model

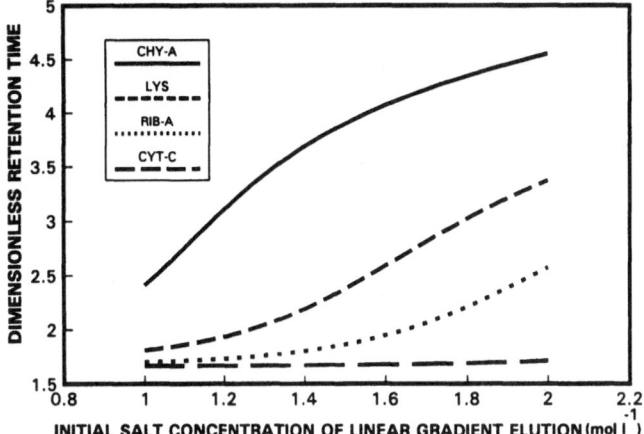

**Fig. 39.** Retention time vs initial salt concentration of linear gradient elution chromatography calculated through the ideal model

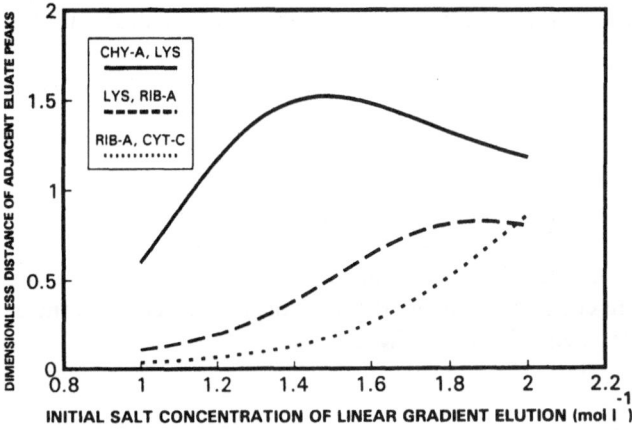

**Fig. 40.** Distance of adjacent peaks vs initial salt concentration of linear gradient elution chromatography calculated through the ideal model

A first guess for the initial eluent concentration is then chosen, so that the ideal peak distances are larger than the dilution ratio and the strongest eluate has the shortest ideal retention time. The second and the third guesses of the initial eluent concentration will be the first one plus and minus a selected small value, respectively. If the separation result with the second guess is better than with the first guess from the calculation through the detailed model, then the fourth guess will be the second guess plus that certain value, and vice versa. The iteration goes on until the separation result satisfies the criteria of separation resolution. After the optimal initial eluent concentration is determined through

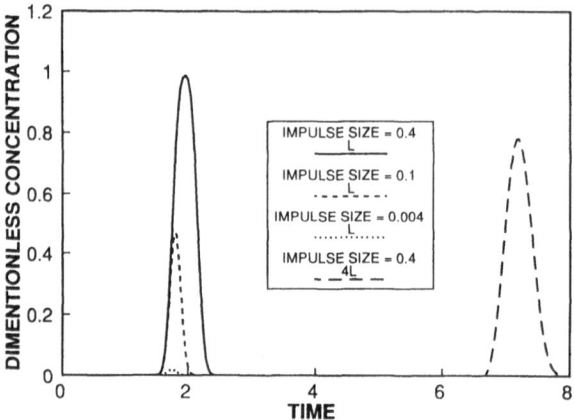

**Fig. 41.** Dilution of feed impulse by pore liquid of porous chromatographic material

the detailed model, the optimal gradient period also can be determined through the same iteration approach as for the optimal initial eluent concentration using the detailed model.

## 5.1 Eluent Concentrations

In this case, 1.7 M and 0.6 M ammonium sulfate are the optimal initial and final eluent concentrations, respectively, as shown in Fig. 24. The separation performance according to the optimal eluent concentrations is compared with that of 1.6 M as the initial eluent concentration and 0.4 M as the final eluent concentration of a 2 min linear gradient, shown in Fig. 25.

## 5.2 Gradient Period

A shorter gradient period of a linear gradient, such as 1 min, can save operation time but lowers the separation performance, as shown in Fig. 28. A longer gradient period, such as 4 or 10 min, can improve the separation performance, but is not time effective, and the bands are broader than in a shorter gradient period, as shown in Figs. 29 and 30. An optimal gradient period, 2 min in this case, can be determined, as shown in Fig. 24.

## 5.3 Flowrate

For the same elution volume of the eluent solution, the flowrate is inversely proportional to the gradient period. Therefore, increasing flowrate will reduce the separation performance, as shown in Figs. 24, 26 and 27, although it can save operation time. Increasing flowrate will also result in a high pressure drop within the gradient system. There is usually a pressure limit for most chromatographic devices.

## 5.4 Column Length

Increasing the column length has long been used as a universal method for improving the separation efficiency [72]. This practice is based on the increase in the distance between the eluate bands due to an increase in column length. However, increasing column length also often broadens the bands. If the increase of the band distance is larger than the increase of the band widths when the column length is increased, the separation performance will be improved. Otherwise, the separation performance can be reduced, as illustrated in Fig. 42 [55].

Longitudinal dispersion, slow film mass transfer, intraparticle diffusion, and adsorption and desorption kinetics, can broaden band profiles [12, 13, 32], and the width of the broadened part of the band is proportional to the time of passage through the column, which is called proportional-pattern behavior [32, 73]. The nonlinear response of the stationary phase at the trailing edge of the band can also broaden it [32]. Thus, the band can be broadened by the increase of the column length, because the time of passage through the column is increased with the column length based on the proportional-pattern behavior.

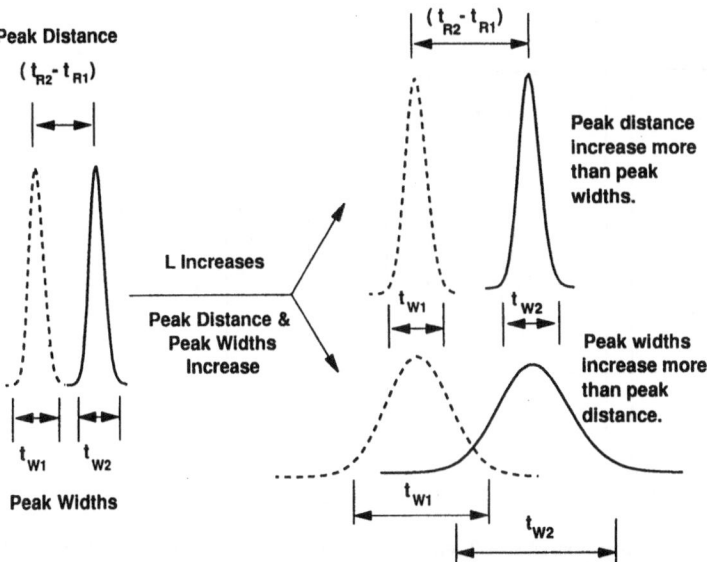

**Fig. 42.** Illustration of the column length effect on the separation performance

On the other hand, the nonlinear response of the stationary phase on the band front is to make it steeper, the so-called self-sharpening effect, and the width of the band front tends towards a constant value as the band moves through the column, called the constant-pattern behavior [32]. Apparently, the constant-pattern behavior improves and the proportional-pattern behavior damages the separation performance. The conflict between the constant-pattern and the proportional-pattern behaviors can result in a mixed outcome of chromatographic separations, instead of other solely constant-pattern behavior or solely proportional-pattern behavior. Thus, increasing column length is expected to improve or damage the separation performance according to constant-pattern or proportional-pattern behavior, respectively. Then, an optimal column length may exist in a system exhibiting both behaviors, as shown in Fig. 43 [55].

The plate theory indicates that the plate number or the separation efficiency is proportional to the column length [27, 74]. In displacement chromatography, Golshan-Shirazi et al. [75], showed that if the sample is smaller than the optimum loading factor, the isotachic train will be formed before the end of the column and increasing the column length will result in no change in the band profiles, which is consistent with constant-pattern behavior. However, the plate theory is limited to symmetric Gaussian bands and linear chromatography. For asymmetric bands, an optimal column length may exist [55].

Furthermore, the dilution of the injection band by the pore liquid also can broaden the band, as shown in Fig. 41 [55]. However, this phenomenon has long been overlooked. A longer column contains more pore liquid, which also can make the injection band more diluted and broader.

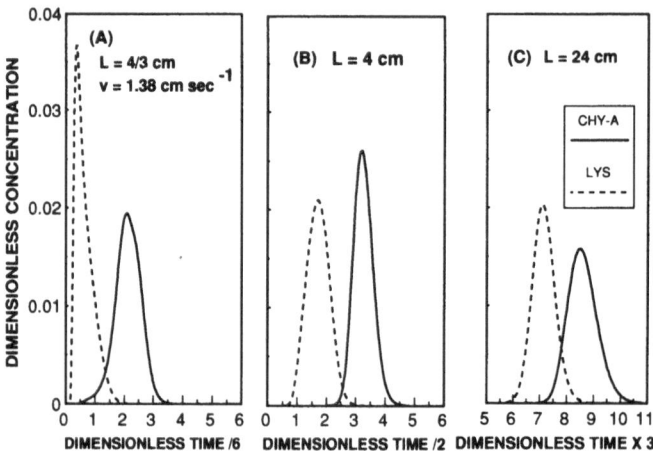

**Fig. 43.** Comparison of chromatograms of 2 min linear gradient elution from 1.7 to 0.6 mol l$^{-1}$ ammonium sulfate at various column lengths calculated through the detailed model. (A) $4 \times 3^{-1}$ cm, (B) 4 cm, (C) 24 cm; v = 1.38 cm s$^{-1}$

## 5.5 Gradient Shape

In comparing linear gradients with stepwise gradients, there is no unique answer to which is better. The ratio of the maximal loading capacity to the operation period can be a standard for the separation performance. However, the stepwise gradient is more cost effective due to its smaller instrument requirements. But, to separate relatively close eluates, the linear gradient usually can achieve better separation relative to the stepwise gradient due to the continual increase of the elution strength throughout the gradient period. Likewise, a similar comparison can be made between linear gradient and isocratic runs. On the other hand, to separate dissimilar eluates, a stepwise gradient could be economical and efficient enough. Furthermore, although nonlinear gradients and segmented linear gradients have the advantage of higher separation efficiency they also have the disadvantage of inconvenient complexity.

## 5.6 Process Tolerance to the Fluctuation of Input Parameters

In industrial operations of chromatographic processes, it is not easy to control the input conditions as precisely as in the laboratory. The particle size of the column material may vary from batch to batch. Pure or reagent grade reagents may not be used in large-scale production. The column capacity may degrade over time [76]. Lot-to-lot consistencies may not be good. Some of the input parameters are uncontrollable such as the concentrations of the bioconversion-generated feeds. Variation of bioconversion potency by 10%–30% among batches is not unusual. The concentrations of some bioconversion-generated trace compounds can vary by hundreds of percentage points. A good separation process must consider not only the separation efficiency but also the process tolerance to the fluctuation of input conditions.

Sometimes, the process tolerance to the fluctuation of input conditions is contradictory to separation efficiency. From the stoichiometric model of reversed phase chromatography, $\log(k')$ ($k'$ denotes the capacity factor) is proportional to $\log(Cm)$ ($Cm$ denotes the eluent concentration) with a proportional coefficient $Z'$ [58]. The $Z'$ values of proteins can be in the order of hundreds. This implies that a 1% deviation of eluent concentration can result in the change of retention time of the eluates up to 1000%. Therefore, this process has a very good separation efficiency due to the large $Z'$ value, but the process stability has a very poor tolerance to fluctuation of $Cm$.

The optimization strategy has been developed by considering both the separation efficiency and the process tolerance to the fluctuation of input conditions, as shown in Fig. 44 [55]. A case in point is the fluctuation of the eluent and the eluate concentrations in stepwise gradient elution chromatography, as illustrated in Figs. 34–36. The worst or largest fluctuation of the input parameters must be defined first, 0.1 M for the second-step eluent concentration in this case; then the optimization strategy can be developed to ensure the

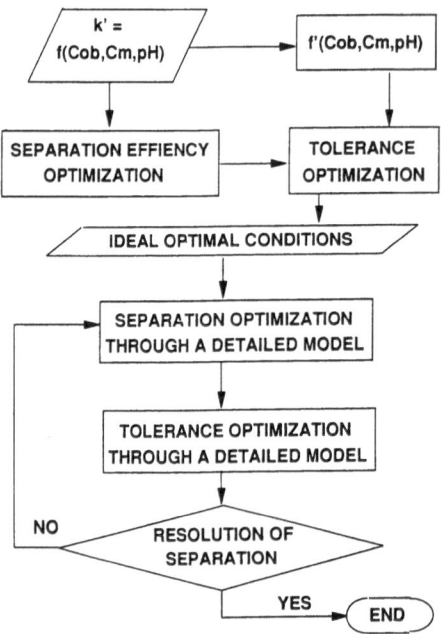

**Fig. 44.** Flow-sheet illustration of optimization strategy of separation efficiency and tolerance to the fluctuation of input parameters

separation performance under the worst operation conditions. The optimal eluent concentration of the second step in this is 1.6 M, as shown in Fig. 34. However, this optimal eluent concentrations has been adjusted to 1.7 M after considering the process tolerance to the fluctuation of the input parameters, as shown in Fig. 35.

*Acknowledgement.* This article is prepared with the grant support (EET-8613167A2) from the National Science Foundation.

# 6 References

1. Jandera P, Churacek J (1985) Gradient elution in column liquid chromatography. Elsevier, Amsterdam
2. Liteau C, Gocan S (1974) Gradient liquid chromatography. John Wiley, New York
3. Schoenmakers PJ (1986) Optimization of chromatographic selectivity. Elsevier, Amsterdam
4. Snyder LR (1980) Gradient elution. In: Horvath C (ed) High-performance liquid chromatography: Advances and prospectives, vol 1. Academic, New York
5. Verzele M, Daweale C, van Dijck J, van Haver D (1982) J Chromatogr 249: 231
6. Knox JH, Pyper HM (1986) J Chromatogr 215: 295
7. Helfferich FG (1962) Ion exchange. McGraw-Hill, New York
8. Frenz J, Cs. Horvath Cs (1985) AIChE J 31: 400
9. Helfferich FG (1986) J Chromatogr 373: 45–60
10. Rhee H-K, Aris R, Amundson NR (1970) Philosophical Transactions Royal Society London 1182: 419

11. Snyder LR (1968) Principles of adsorption chromatography. Marcel Dekker, New York
12. Chase HA (1984) C.E.S. 39: 1099
13. Muller AJ, Carr P (1984) J Chromatogr 294: 235
14. Guiochon G, Ghodbane S, Golshan-Shirazi S, Huang J-X, Katti A, Lin B-C, Ma Z (1989) Talanta. 36: 19
15. Arnold FH, Blanch HW (1986) J Chromatogr 355: 13
16. Arve BH, Liapis AI (1987) AIChE J 33: 179
17. Biyani P, Goochee CF (1988) AIChE J 34: 1747
18. Cen PL, Yang RT (1986) AIChE J 32: 1635
19. Huang CC, Fair JR (1984) AIChE J 34: 1861
20. Seshadri S, Deming SN (1984) Anal Chem 56: 1567
21. Tsou H-S, Graham EE (1985) AIChE J 31: 1959
22. Van Vuuren DS, Stander CM, Glasser D (1984) AIChE J 30: 593
23. Carta G (1988) C.E.S. 43: 2877
24. Furusaki, Haruguchi E, Nozawa T (1987) Bioprocess Engineering 2: 45
25. Horvath Cs, Lin H-J (1976) J Chromatogr 126: 401
26. Horvath Cs, Lin H-J (1978) J Chromatogr 149: 43
27. Snyder LR, Kirkland JJ (1979) Introduction to modern liquid chromatography. Wiley, New York
28. Guiochon G, Golshan-Shirazi S, Jaulmes A (1988) Anal Chem 60: 1865
29. Klein G (1985) AIChE Symposium Series, 242: 28
30. Gu T, Tsao GT, Tsai G-J, Ladisch MR (1990) AIChE J 36: 1156
31. Ruthven DM (1984) Principles of adsorption and adsorption processes. John Wiley, New York
32. Sherwood TK, Pigford RL, Wilke RW (1975) Mass transfer. McGraw-Hill, New York
33. Tokieda T, Tokuda T, Ishida M (1985) Anal Sci 1: 395
34. Jandera P, Churacek J, Colin H (1981) J Chromatogr 214: 35
35. Klotz IM (1946) Archives Biochem 9: 109
36. Mazsaroff I, Varady L, Mouchawar GA, Regnier FE (1990) J Chromatogr 499: 63
37. Melander WR, El Rassi Z, Horvath Cs (1989) J Chromatogr 469: 3
38. Srinivasan R, Ruckenstein E (1980) Separation purification methods 9: 267
39. Melander WR, Horvath Cs (1977) Archives Biochem Biophys 183: 200
40. Roettger BF, Myers JA, Ladisch MR, Regnier FE (1989) Biotechnol Progress 5: 79
41. Yamamoto S, Nomura M, Sano Y (1987) AIChE J 33: 1426
42. Antia FD, Horvath Cs (1989) J Chromatogr 484: 1
43. Ghrist BFD, Snyder LR (1988) J Chromatogr 459: 25
44. Dolan JW (1988) LC-GC 6: 572
45. Weinberger R, Coniglione V (1984) LC 2: 10
46. Berridge JC (1985) Techniques for the automated optimization of HPLC separations. John Wiley, New York
47. Huber JFK (1978) Instrumentation for high-performance liquid chromatography. Elsevier, Amsterdam
48. MacDonald JC (1986) HPLC: Instrumentation and applications. International Scientific Communications, Fairfield, Connecticut
49. Billiet HA, Keehnen PD, de Galan L (1979) J Chromatogr 185: 515
50. Erni, Frei RW (1978) Res Dev 149: 56
51. Martin M, Guiochon G (1978) In: Huber JFK (ed) Instrumentation for high-performance liquid chromatography, Chapter 3. Elsevier, Amsterdam
52. Pao RHF (1961) Fluid mechanics. John Wiley, New York
53. Scott RP (197 ) J Chromatogr Sci 9: 385
54. Bio-Rad Laboratories Bio-Rad Model 385 Gradient Former Instruction Manual. Richmond, California
55. Truei YH (1991) PhD Dissertation. Purdue University, West Lafayette, Indiana
56. Skidmore GL, Hormann BJ, Chase HA (1990) J Chromatogr 498: 113
57. Heinitz ML, Kennedy L, Kopaciewicz W, Regnier FE (1988) J Chromatogr 443: 173
58. Geng X, Regnier FE (1984) J Chromatogr 296: 15
59. Kopaciewicz W, Regnier FE (1983) Anal Chem 133: 251
60. Pitt WW Jr (1976) J Chromatogr Sci 14: 396
61. Aguilar MI, Hodder AN, Hearn TW (1985) J Chromatogr 327: 115
62. Hearn MTW, Grego B (1983) J Chromatogr 255: 125
63. Mazsaroff I, Cook S, Regnier FE (1988) J Chromatogr 443: 119

64. Jessissen HP (1976) Hoppe Seler's Z Physiol Chem 357: 1727
65. Afeyan NB, Gordon NF, Mazsaroff J, Varady L, Fulton SP, Yang YB, Regnier FE (1985) J Chromatogr 519: 1
66. Yau WW, Kirkland JJ, Bly DD (1979) Modern size exclusion liquid chromatography. Wiley, New York
67. Chung SF, aWen CY (1968) AIChE J 14: 857
68. Wakao N, Oshima T, Yagi S (1958) Kagaku Kagaku 22: 780
69. Cantor and Schimmel, 1980
70. Cooper AR, Barrall II EM (1973) J Applied Polymer Sci 17: 1253
71. Regnier FE, Mazsaroff I (1987) Biotechnol Progress 3: 22
72. Katti AM, Guiochon (1988) J Chromatogr 449: 25
73. Wankat PC, Koo Y-M (1988) AIChE J 34: 1006
74. Gibbs SJ, Lightfoot EN (1986) I&EC Fundam 25: 490
75. Golshan-Shirazi G, Lin B, Guiochon G (1989) Anal Chem 61: 1960
76. Tice PA, Mazsaroff I, Lin NT, Regnier FE (1987) J Chromatogr 410: 43
77. Jandera P, Churacek J (1980) J Chromatogr 192: 19
78. Armstrong DW, Boehm RE (1984) J Chromatogr Sci 22: 378
79. Kennedy L, Kopaciewicz W, Regnier FE (1986) J Chromatogr 359: 73
80. D'Agostino G, O'Hare MJ, Mitchell F, Salomon T, Verllon F (1988) Chromatographia 25: 343
81. Markowski W, Golkiewicz W (1988) Chromatographia 25: 339
82. Martin M (1988) J Liquid Chromatogr 11: 1809
83. Merengo E, Gennaro MC, Baiocchi C, Bertolo P (1988) Chromatographia 25: 413
84. Kang K, McCoy BJ (1989) Biotechnol Bioeng 33: 786

# Synthetic Membranes in Biotechnology: Realities and Possibilities

C. A. Heath[1] and G. Belfort[2]

[1] Biotechnology and Biochemical Engineering Program, Thayer School of Engineering, Dartmouth College, Hanover, New Hampshire 03755-8000, USA

[2] Bioseparations Research Center, Howard P. Isermann Dept. of Chemical Eng., Rensselaer Polytechnic Institute, Troy, New York 12180-3590, USA

Synthetic membrane processes are being increasingly integrated into existing reaction, isolation and recovery schemes for the production of valuable biological molecules. In many cases they are also replacing traditional unit processes. The properties of membrane systems which are most often exploited for both upstream and downstream processing are their high surface area per unit volume, their permselectivity, and their potential for controlling the level of contact and/or mixing between two separate phases. Advances in both membrane materials and module design/operation have led to better control of concentration polarization and membrane fouling. While this article begins by noting some of the recent advances in membrane technology such as new developments in membrane materials and fluid mechanics, followed by integration of membranes into both cellular and enzymatic reaction systems, the primary focus is a review of established and emerging membrane separation processes. Many examples referred to in this review underscore the potential for combining membranes and biological processes to good advantage. We believe that this marriage has only just begun and that improvements in membrane materials and a better understanding of the fluid mechanics in membrane modules and of metabolic processes for reaction systems will lead to an even stronger synergism in the future.

Advances in Biochemical Engineering
Biotechnology, Vol. 47
Managing Editor:
© Springer-Verlag Berlin Heidelberg 1992

# 1  Introduction

## 1.1  Why Integrate Membranes with Bioprocesses?

Concomitant developments in molecular and cell biology and in separation technology during the past thirty years have produced exciting new opportunities in the production of complex mammalian proteins that have the potential to fundamentally alter human healthcare in such areas as diagnostics, prevention, and treatment of disease. One particular class of separation techniques, synthetic membrane processes, is beginning to play an increasingly important role in many aspects of bioprocessing. Not only have membrane processes been used in established separation schemes such as for clarification and concentration and for purification of macromolecular products, but they are also being used in new emerging schemes for the separation and purification of macro and microsolutes. Well known membrane processes such as electrodialysis, reverse osmosis and pervaporation are finding niches in downstream processing of bioconversion and cell culture products. Membrane structures are also being integrated into the bioreactor itself in order to increase volumetric productivity and to reduce subsequent recovery requirements. In addition, membranes are being utilized not only for their permselective properties but also for their large internal adsorptive surface areas and attractive mass transfer characteristics. Microporous membranes provide excellent matrices for group and biospecific adsorption processes such as ion exchange and affinity-based separations, respectively, and for entrapping enzymes. Synergistic advantages of coupling membranes with other unit processes such as affinity ligand adsorption, precipitation, and solvent extraction are also becoming apparent.

Several limitations associated with bioprocessing can be overcome or at least minimized with membrane processes. These include the generation of complex mixtures requiring extensive downstream processing, the production of dilute solutions containing low product concentrations when using suspension cultures, low specific rate constants for biological processes, and contamination and infection of unwanted biological species (especially for mammalian and plant cell cultures). Many of the applications of and recent advances in synthetic membrane technology described in this paper deal with mitigating these limitations. Because of their ability to fractionate and concentrate, membranes are able to separate product from reactant, to increase the concentration of a dilute solution, to reduce the possibility of contamination and infection, and, because they are often closed processes, to reduce the need for stringent contamination requirements. Membranes can also be used to increase the concentration of the biocatalyst and, hence, increase the volumetric productivity of a reactor.

## 1.2  Recent Advances in Membrane Technology

### 1.2.1  Novel Membrane Materials and Structures

A major challenge facing membrane chemists has always been the need to relate chemical structure with membrane function. An example of progress in this area has been the recent work in understanding the chemical structure and performance of gas permeable glassy polymers such as poly(trimethyl silyl 1-propyne; PTMSP) at room temperature. The chemical structure of this glassy polymer is shown in Fig. 1a, where it can be seen that the fairly rigid acetylene backbone gives the polymer stability while the large trimethyl-silicon groups impart a very large free volume between adjacent acetylene backbone chains. It is this property that researchers feel give PTMSP its exceptionally high transport characteristics. Several workers have attempted to increase the free volume further and to change the polarity of the lone $CH_3$ group by replacing it with

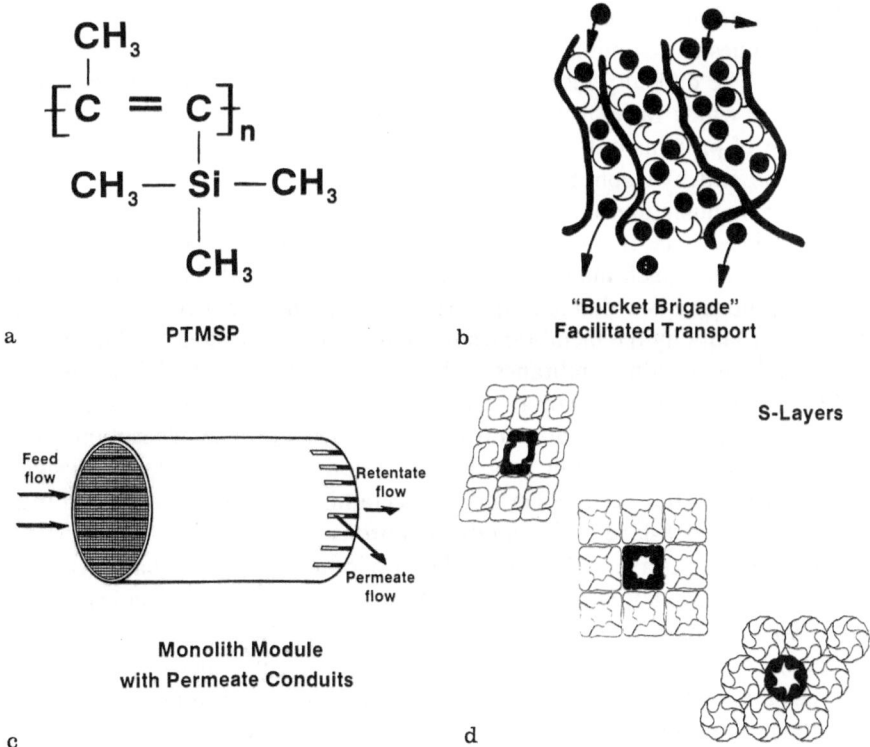

**Fig. 1a–d.** Novel membrane materials and structures. **a**) Chemical structure of poly (trimethyl silyl 1-propyne). **b**) By covalently attaching affinity or complexing agents to the membrane pore surfaces, the selectivity of membrane pore processes can be increased as molecules are adsorbed onto the surface. **c**) Ceramic manifold (after Goldsmith [9]). **d**) S-layers (after Sara and Sleytr [10])

poly(dimethyl siloxane; PDMS) [1]. Others have replaced the hydrogen molecules with fluorine molecules in an attempt to change the polarity of the polymer and increase its selectivity [2]. These efforts have increased the methanol-water separation factor for pervaporation from a value of about 12–14 for the base polymer of the PTMSP to a value of about 40 for these modified polymers [3].

Another dream of polymer chemists has been to try to take advantage of the highly selective affinity between certain gases or solute molecules and complexing agents. The idea is to suspend the complexing agents, which have high affinity for the particular solute that is to be transferred across the membrane, in a solvent within the capillary structure of the membrane. Although this is an exciting area of research, industrial enthusiasm for this idea has not been overwhelming. Another variant of this approach is to covalently attach the complexing or affinity agent to the membrane pore surface. A schematic of this idea is shown in Fig. 1b in which the permeable molecule jumps onto its polymeric ligand and then moves along by surface or pore diffusion to the next ligand, and to the next, etc. One example of this process is the use of crown ethers covalently bound to a matrix in order to selectively transport oxygen in preference to nitrogen through a membrane.

Concentration and fractionation of proteins by membrane filtration, in some cases, continues to be limited by fouling despite advances in fouling prevention techniques. The difficulty rests in an incomplete understanding of fouling, the behaviour of which varies from one process to another. The general view is that adsorption of proteins onto membrane surfaces (within pores) is governed by electrostatic and hydrophobic interactions. Because most synthetic polymers used in the manufacture of membranes are relatively hydrophobic in character they must be surface treated, frequently with hydrophilic macromolecules or ionic surfactants, which may reduce fouling by proteins [4]. Manufactures are currently following two different approaches to minimizing membrane fouling. The first is to modify the membrane surface to minimize protein binding which is commonly done with membranes such as polysulfone, polyvinylidene difluoride (PVDF) and polytetrafluorethylene (PTFE). The second is to make membranes from materials which do not result in significant protein adsorption. New materials with low non-specific protein adsorption or narrow molecular weight cut-off characteristics such as asymmetric ceramic microfiltration membranes [5], recrystallized S-layers [6], modified porous glass [7], and polymerized Langmuir-Blodgett layers [8] are in the development stage or have recently been commercialized. Figures 1c and 1d show schematics of a ceramic manifold and of S-layers, respectively.

Ceramics are becoming an attractive alternative to polymers as membrane materials due to their excellent chemical, thermal, and mechanical properties. Advantages for biotechnological processes include ease of cleaning due to chemical stability, steam sterilizability, sharp pore size distribution, and high fluxes because of the ability to withstand higher pressures [5]. These qualities make ceramic membranes good candidates for immobilized whole cell reactors, cell harvesting, and protein purification.

Two relatively new commercial ceramic membranes have recently been developed in the United Kingdom. These include the electrochemically produced ceramic membranes with very ordered pores and the new stainless steel screen reinforced ceramic membrane shown in Fig. 2. To create the former, an aluminum metal substrate is anodized in an electrolyte such as sulfuric or phosphoric acid, and an anodic oxide film is formed on the surface [11]. This film has a relatively thick porous layer comprised of regularly spaced pores extending from the outer surface in toward the metal. As the anodizing continues, metal is converted to oxide at the metal oxide interface and the pores extend further into the film. The thickness of the barrier layer remains relatively constant. The cross section and spacing of the pores and the thickness of the barrier layer are all proportional to the anodizing voltage. It is possible to remove the porous oxidizing layer (anodic oxide film) by dissolution in acid or alkaline. The remaining porous ceramic film or membrane can be used as a filter. This technology has been used to produce the membrane shown in Fig. 2a

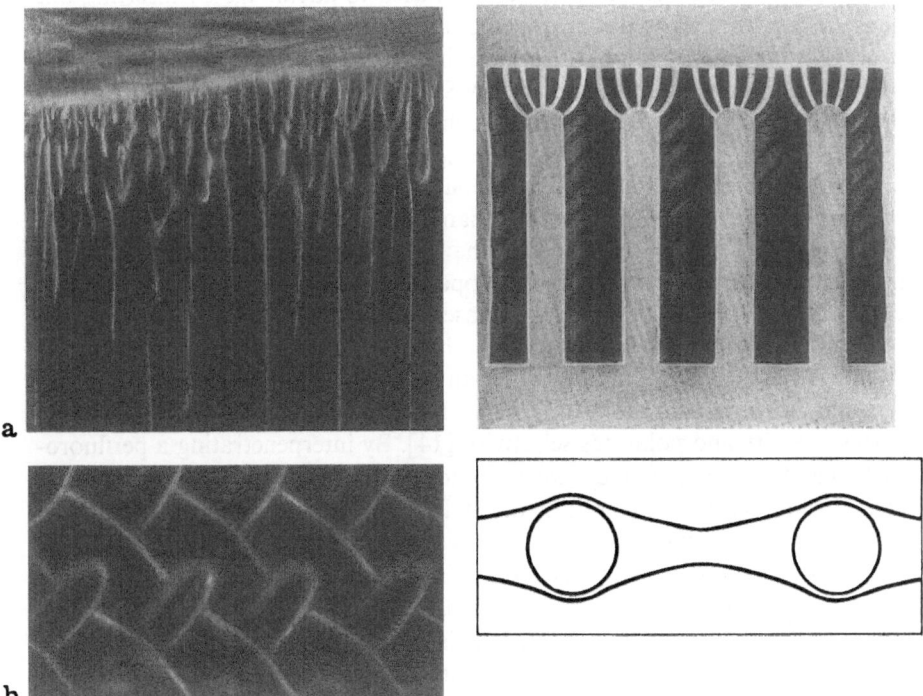

**Fig. 2a, b.** Two different commercial ceramic membranes. a) Electrochemically-produced ceramic membrane showing the uniform pores in the top skin layer and in the support structure. b) Ceramic membrane reinforced by a metal screen showing its rough surface

(ANOPORE, Anotec Separations, Banbury, UK), which has very sharply defined, straight pores and a pore density of 50%. The membrane is also non-toxic, low protein-binding, and transparent when wet.

Metal reinforced ceramic membranes (e.g. from Ceramesh Ltd., Banbury, UK), are made from partially sintered microporous ceramic suspended in the apertures of a metal based superalloy mesh [12]. The result is a highly formable and flexible membrane with a narrow pore size distribution and high chemical and thermal resistance. The dimpled surface, which is only about 20 μm thick in the meniscus-like cross sections of the ceramic septa, is expected to affect the hydrodynamics at the interface (Fig. 2b). The actual effects of this surface topography have not yet been determined.

In comparison to most ceramic membranes, porous glass membranes [7] are homogeneous and generally symmetric. Pore diameters can be as small as a few nanometers and the surface properties of glass can add an additional means of separation. The glass surface can be hydrophilic or hydrophobic with different functional groups and/or affinity ligands. These membranes (Bioran, Schott, Mainz) take advantage of the dual filtration and chromatographic effects. The capillary membranes are typically 50–60% porous and can have different membrane thicknesses.

A relatively old idea has recently been used to advantage by inserting one phase within another to obtain a membrane with separate phases and different properties. Hennepe et al. [13] have inserted silicalite (5.4 μm pores and $0.20 \, cm^3 \, g^{-1}$ pore volume) into a PDMS polymer film. The silicalite particles drastically change the relative transport characteristics of ethanol and water in a pervaporation membrane process resulting in an increased separation factor for preferential ethanol transport from about 12 for PDMS to about 38 for the composite membrane (Fig. 3a). The ethanol flux also increased twofold.

A relatively new development is the formation of a membrane film with interpenetrating contiguous phases. Apparently, unusual electronic and ionic conductive properties result as one phase constrains the behavior of a second phase (Figs. 3b and 3c). For example, a hydrophobic phase can constrain the swelling due to water imbibition of an ionic hydrophilic electrolyte phase resulting in a membrane with unusually high ionic conductivity, cationic permselectivity and polar gas selectivity [14]. By interpenetrating a perfluoro-carbon matrix (hydrophobic) with a second contiguous phase of highly sulphon-ated perfluorocarbon material (hydrophilic), the former matrix constrains the swelling of the latter matrix, thus preventing the hydrophilic matrix from absorbing large amounts of water when immersed in aqueous electrolyte. The result of this constraint is a membrane (i.e. Nafion, DuPont) that has high ionic density when saturated with water. This membrane displays superior ionic conductivity, either in dilute or concentrated electrolyte solutions, and ex-tremely high cationic permselectivity. Michaels [14] has suggested that such a membrane could be expected to function as a selective gas permeation mem-brane with high permselectivity and permeability to polar gases through the concentrated aqueous electrolyte solution. Actually, Way et al. [15] have

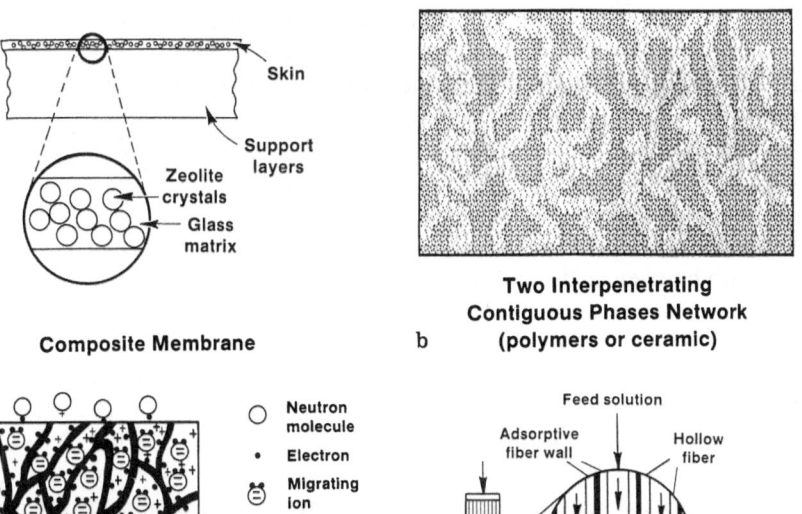

**a    Composite Membrane**

**b    Two Interpenetrating Contiguous Phases Network (polymers or ceramic)**

**c    Electrochemically Permselective Heterogeneous Membrane**

**d    Hollow-fiber Chromatographic Column**

**Fig. 3a–d.** Novel membrane materials and structure. **a)** Composite membrane showing zeolite crystals (or other particles) distributed in the skin layer. **b)** Membrane with two interpenetrating contiguous phases. **c)** Electrochemically permselective heterogeneous membrane. **d)** Hollow fiber chromatographic column (after Michaels [14])

confirmed this with the separation of carbon dioxide from methane using amine-containing perfluorosulfonic acid ion exchange membranes.

Using a similar approach, one can construct an electrochemically permselective heterogeneous membrane as shown in Fig. 3c. In this concept an electronically conductive (i.e. metal or semi-conductor) phase is interpenetrated with an ionically conductive (i.e. polyelectrolyte or ceramic solid electrolyte) phase. In this system, electrons conduct through the metallic or semi-conductive phase and ions diffuse through the electrolyte phase. Should a gas species entering one of the faces be capable of releasing its electrons to the electronically conductive phase, it will then move through the membrane as an ionic species to the second face. At this face, it will recombine with the electrons from the metallic phase and leave as an uncharged species at the downstream membrane boundary. Michaels [14] suggests that a membrane comprised of platinum and zirconia, for example, might function as an oxygen permselective membrane at elevated temperatures. Another membrane composed of platinum and hydrated polystyrene sulphonic acid could act as a hydrogen selective membrane. In both cases, microporous ceramic matrices could be used to great effect.

Recent developments have also led to the use of hollow fiber membranes as a chromatographic matrix. This is shown in Fig. 3d where one can take advantage of the extremely well-defined fluid mechanics in such systems [16]. This is discussed in more detail in Sect. 5.1.

Another exciting and very recent development was reported by Anderson et al. [17]. They were able to tailor the gas selectivity and permeability of a conjugated polyaniline $[(C_6H_4NH)_x]$ membrane by doping, undoping, and redoping the polymer film with counterions of appropriate size. The doping process with halogen acids made these materials conductive and allowed precise changes to be made in their morphology. Undoping increased the membrane permeability to small gases relative to that of the as-cast film. Partial redoping then allowed controlled changes to the membrane free volume resulting in exceptionally high permeabilities and separation factors. For example, the selectivity values of 3590 for $H_2/N_2$, 30 for $O_2/N_2$, and 336 for $CO_2/CH_4$ surpass the highest previously reported values of 313, 16, and 60, respectively.

Another interesting and potentially widely practical development is the new "particle membrane" developed by Louis Errede, Don Hagen and others (Empore, 3M, St. Paul, MN) [18]. A suspension of porous particles is mixed with a suspension of polytetrafluoroethene (PTFE, Teflon) and squeezed between roller drums into a film containing the original particles held together by fibrillated PTFE. Preparation of membranes with widely differing properties is thus possible depending on the choice of the initial particles. By using a wide array of particles or beads such as those developed for column chromatography, adsorptive membranes can easily be prepared.

### 1.2.2 Fluid Management in Membrane Module Development

Typical commercial membrane modules were originally developed to treat aqueous solutions essentially free of macromolecules (proteins) and suspended particulates. Most manufacturers have, however, adapted these early hollow fiber and spiral wound module designs for treating biological suspensions containing proteins, cells and cellular debris. Unexpected and severe fouling problems have occurred as a result of protein adsorption and deposition and buildup of particulate matter at the membrane-solution interface. Methods to alleviate these limitations include the development of low non-specific protein adsorbing materials (as discussed above) and the design of new modules that induce flow instabilities that could reentrain the so-called "cake" on the membrane surface back into the flowing solution. Modules with furrowed membrane surfaces, with membranes on the faces of an annulus between a rotating inner cylinder and a stationary outer cylinder, and with a spiral half cylinder cross-sectional channel placed on a flat sheet membrane have appeared on the market [19–26]. All of these designs use flow instabilities such as vortices to sweep the surface of the membranes (Fig. 4). Pulsating axial flow and flow backwashing have also been used to clean membrane surfaces and blocked

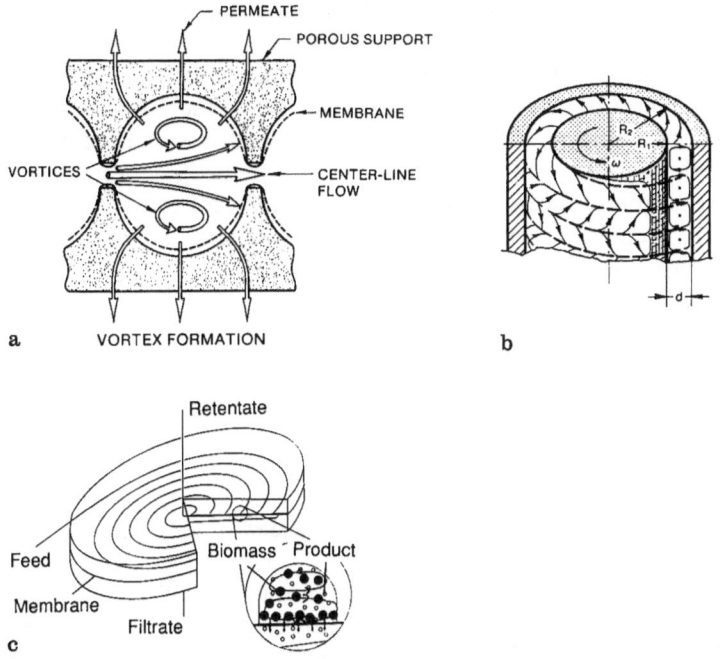

**Fig. 4a–c.** Membrane module designs that use flow instabilities (vortices) to sweep the surface of the membranes. **a)** Corrugated surface showing vortex formation (after Edwards and Wilkinson [23]). **b)** Rotating inner surface and stationary outer surface resulting in the formation of Taylor vortices (from Sobey [24]). The membrane can be placed onto both or either surface. **c)** Spiral half cylinder channel showing secondary centrifugal flow with formation of Dean vortices (after Winzeler [25])

pores [27]. Localized instabilities at the solution-membrane interface can also be induced by vigorously vibrating the membrane surface (i.e. by oscillating cylindrical or leaf membrane elements) resulting in substantially increased wall shear rates. According to Culkin [28], most of the input energy is localized at the mass boundary layer thereby depolarizing concentration polarization and fouling. Culkin reports permeation rates 5 to 10 times higher with this system than with crossflow systems. Unfortunately, to date very few data have been published in a refereed journal and the limitations of this approach such as sealing and difficulty in scale-up have not been adequately addressed.

By solving the Navier-Stokes equations for the trajectories of particles in dilute suspensions in laminar flow in a porous slit or tube, Belfort et al. [29] have provided a module design rationale (and procedure) for capture or non-capture of suspended particles. Thus, within the context of the assumptions of this approach, membrane modules can be designed that minimize capture of particles by membranes and hence are less susceptible to particulate fouling.

Integrating membranes into bacterial and mammalian cell culture bioreactors and into enzyme reactors are the topics covered in the next section of this paper. Because of their ability to compartmentalize, membranes are finding use as barriers to separate cells and the flowing culture medium or to separate product from the cells in a variety of culture systems. The high surface area per unit volume also makes them excellent candidates as matrices for immobilizing or entrapping enzymes. Included is a brief discussion of the advantages for utilizing membranes and a review of some commercial devices and innovative designs still in the development stage. On the downstream side, established membrane bioseparation processes will be covered briefly, followed by a more in-depth coverage of emerging applications including integration of membranes with other unit processes. New opportunities for and future developments in membrane processes will be discussed in the final section.

# 2 Membrane Bioreactors

## 2.1 Integration of Membranes into Bioprocesses

Biotechnology is based on the use of living organisms and enzymes for the manufacture of commercial products and is currently exemplified by recombinant DNA and monoclonal antibody technologies. Besides the bioreactor in which bacterial, yeast, mammalian or other cells are used to convert raw materials and medium to final product, the manufacturing process is similar to those used for the manufacture of small organic drugs and other traditional pharmaceutical products (Fig. 5; [30]). In classical bioconversion and recovery processes, there are many opportunities for using synthetic membranes as shown in Fig. 6. Membranes have found use in filtering the raw materials (including gases) and medium prior to entering the bioreactor. In conjunction with the bioreactor, membranes may be used for cell retention and for removal of inhibitory products. During downstream processing and recovery, microfiltration has been used for cell harvesting and for removal of particulates such as cell debris. For primary isolation, membrane processes have also been used together with solvent extraction, coupled transport, and affinity processes. Finally, during the purification step, membranes have been used as porous supports for adsorption [31] and for electrodialysis [32].

## 2.2 Immobilized Whole Cell Reactors

Well-established membrane bioreactors will not be discussed in great depth since several recent reviews are available [33–37]. However, some of the advantages and limitations of microcapsules, hollow fiber, and flat sheet reactors will be mentioned since there is growing interest in using these systems for culturing cells. New and innovative membrane bioreactors that have specific advantages will be described in more detail.

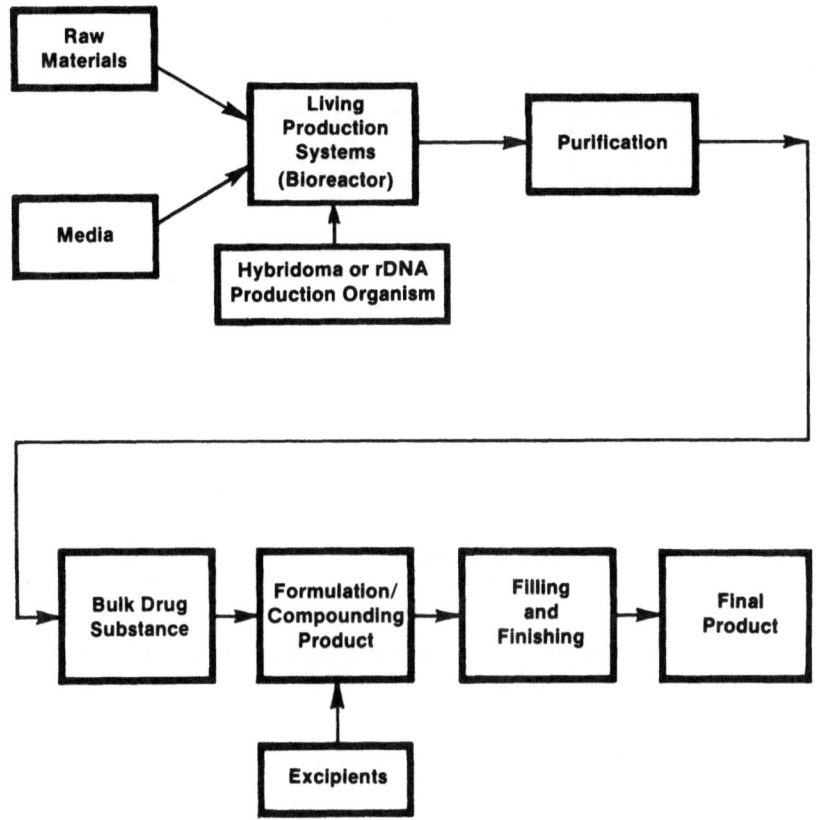

**Fig. 5.** Typical manufacturing process for the production of recombinant DNA-derived biologicals or monoclonal antibody products [30]

The use of microcapsules for mammalian cell growth has been studied by several groups for both biotechnological [38–40] and biomedical [41, 42] purposes. The technology was commercialized several years ago for mammalian cell culture (Abbott Biotech, Needham, MA). In recent years, the concept has also been used for immobilizing enzymes as discussed in Sect. 3.1. Typically, a rich dense culture of hybridomas in microcapsules can be obtained within two weeks. Unfortunately, when the cell density becomes very high, the cells begin to lose viability because of diffusion limitations of either the waste products out of the microcapsule or medium nutrients into the capsule. These diffusion limitations have been analyzed quantitatively by Heath and Belfort [43]. The results shown in Fig. 7 demonstrate that a capsule of 100 μm diameter does not have a concentration gradient across the intracellular mass while a capsule of 500 μm diameter has a severe radial gradient for the particular conditions indicated. The cells become oxygen limited at a distance of less than 20% of the dimensionless radius from the outer wall. Controlling capsule size is of utmost importance in

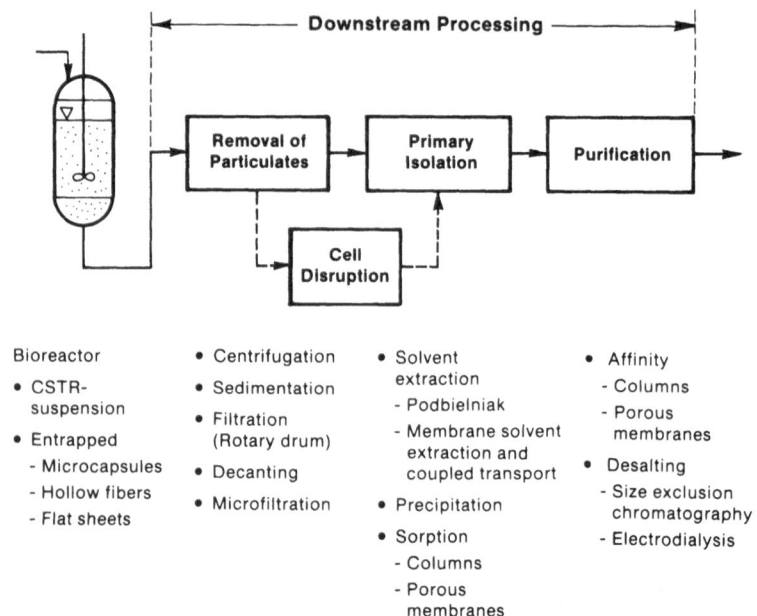

**Fig. 6.** Classical processing/bioconversion and recovery processes showing different unit processes and illustrating the opportunities for synthetic membrane processes

avoiding mass transfer limitations. Other important design parameters include membrane thickness and porosity.

Hollow fiber modules have also been used successfully for culturing mammalian cells. Cells are usually grown in the extracapillary space with medium flow through the fibers but they have also been grown within the fibers with medium flow outside or across the fibers. Growth of hybridomas in the shell space of hollow fiber reactors for the case of mass transfer in the shell space by diffusion only has also been quantitatively analyzed by Heath and Belfort [43] as shown in Fig. 8. In the sample simulation (Fig. 8a), along the axial flow path, oxygen becomes limited at about 20% of the axial distance from the inlet under the simulation conditions indicated. Radial profiles of oxygen shown in Fig. 8b indicate that cells only a short distance from the fiber (20% of the dimensionless radial cell region thickness) are without oxygen at all three axial positions. The system kinetics and mass transfer characteristics will significantly affect the concentration profiles in the reactor. The importance of even, well-defined spacing between the fibers should be emphasized as a means of reducing these mass transfer limitations.

A method of obviating the potential problems mentioned above for conventional (randomly potted) hollow fiber membranes is to insert one hollow fiber within another hollow fiber and to grow the cells in the annulus between the two membranes. By careful choice of the diameters of the inner and outer fibers, one can limit the distance needed for diffusion of medium components, i.e., oxygen,

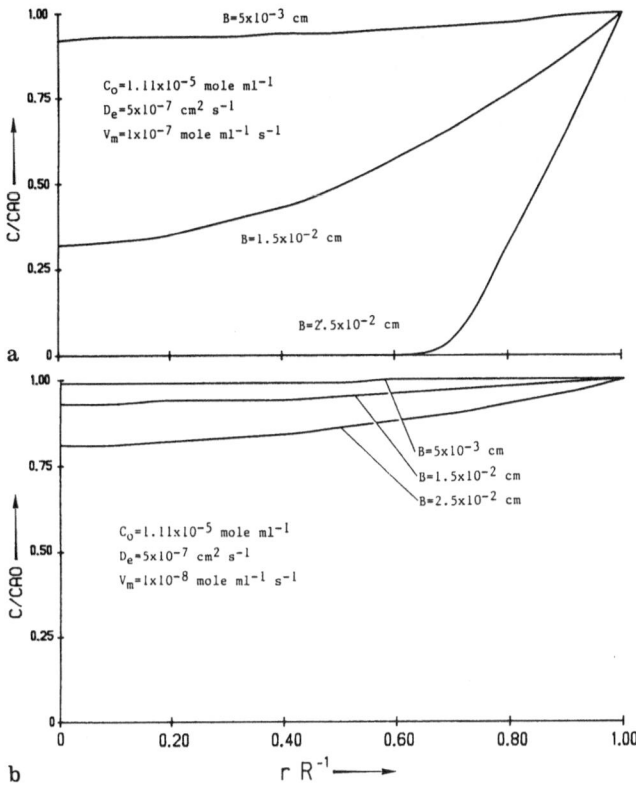

**Fig. 7a, b.** Radial concentration profiles for glucose in microcapsules assuming a zero-order kinetic limit for substrate consumption. The maximum reaction rates are **a)** $1 \times 10^{-7}$ mol ml$^{-1}$ s$^{-1}$ and **b)** $1 \times 10^{-8}$ mol ml$^{-1}$ s$^{-1}$ where B is microcapsule radius, $C_o$ is bulk glucose concentration, $D_e$ is effective diffusivity, and $V_m$ is the substrate consumption rate (from Heath and Belfort [43])

glucose, glutamine and other nutrients, to approximately 50 microns for the furthest cells. The medium is pumped both within the lumen of the inner fiber and along the outside of the outer fiber (Fig. 9). A comparison of the conventional and concentric reactors producing antibodies from hybridoma cells is shown in Table 1 in which it can be seen that much higher concentrations of cells can be maintained in the concentric reactor along with higher specific antibody productivity and cell viability [44].

Another method of maintaining a constant well-defined cell region thickness is the use of flat sheet reactors which has been reported by Rainen [45]. A diagram of their flat sheet system is shown in Fig. 10. Although multi-layer flat sheet membrane bioreactors have lower surface area to volume ratios than hollow fiber systems, they do possess the other advantages of hollow fiber reactors and even overcome some of the disadvantages. Controlled cell region thickness is one such extra advantage; the potential for cell replacement with appropriate design of the device is another.

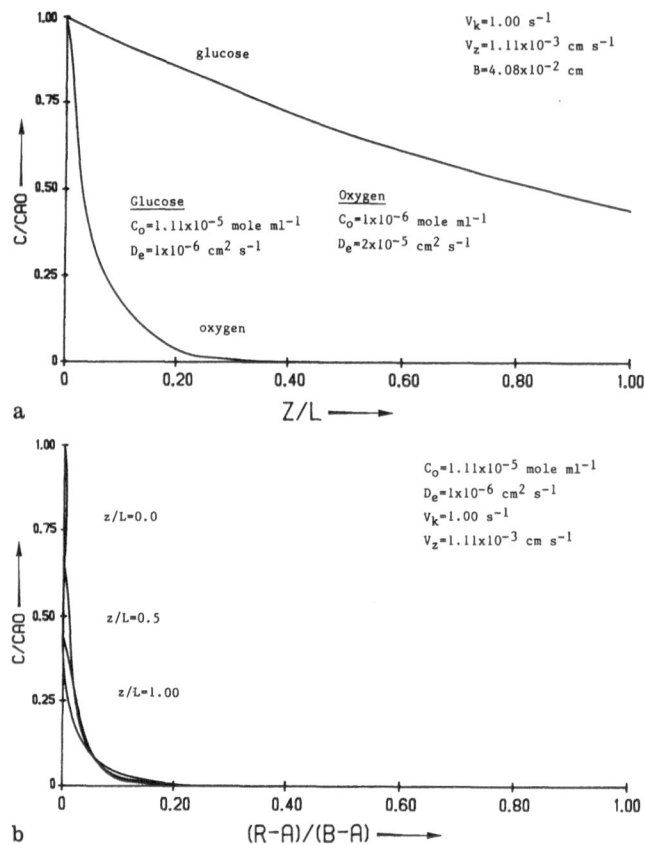

**Fig. 8 a, b.** Concentration profiles for oxygen and glucose in hollow fiber reactors assuming a first-order kinetic limit for substrate consumption: **a)** axial concentration gradients of oxygen and glucose; **b)** radial concentration profiles for glucose where B is radius of a single fiber unit, $C_o$ is bulk concentration, $D_e$ is effective diffusivity, $V_k$ is the maximum reaction rate divided by the Michaelis constant, and $V_z$ is the axial fluid velocity in the lumen (from Heath and Belfort [43])

Tharakan and Chau [46] have cultured mammalian cells in a hollow fiber bioreactor by running the system under pure radial convective flow in an attempt to alleviate detrimental nutrient and other metabolic gradients in the extracapillary space. The process is reminiscent of the hollow fiber perfusion reactor described by Feder and Tolbert several years ago [47]. The Tharakan and Chau reactor consists of a central flow distributor tube containing medium surrounded by an annular bed of hollow fibers containing air and carbon dioxide. The cells are grown in the space outside the fibers. The medium is forced from the distributor tube through the cell bed into the gas-containing fibers, from which it leaves the reactor along with the flowing gases. Concentrated product is periodically bled from the shell space. The same group has also

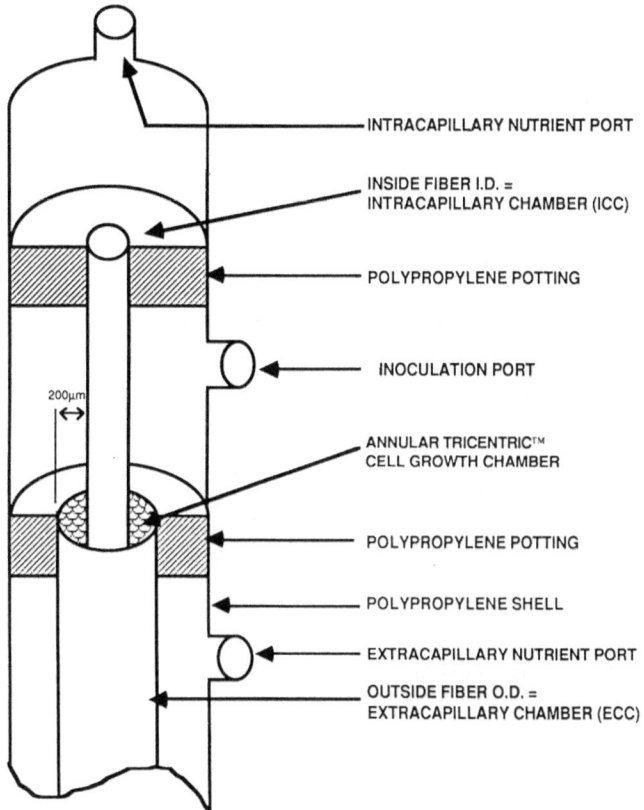

**Fig. 9.** Tricentric bioreactor showing the annular region into which the cells are placed as a result of inserting one hollow fiber into another (courtesy of Bob Boice, Setec, CA)

**Table 1.** Comparison of conventional and concentric reactors for producing antibodies[1]

|                                                              | Conventional Reactor | Concentric Reactor            |
| ------------------------------------------------------------ | -------------------- | ----------------------------- |
| Maximum viable cell density $(ml^{-1})$                      | $2 \times 10^7$      | $1.5–1.8 \times 10^8$         |
| Support true maintenance                                     | no                   | yes                           |
| Diffusional distances $(\mu m)$                              | 400–2000             | 100                           |
| Specific antibody productivity $(\mu g\,(10^9\,cells)^{-1}\,h^{-1})$ | 200–400              | 600–900                       |
| Cell viability (%)                                           | 15–30                | 80–90                         |

[1] From Custer [44]

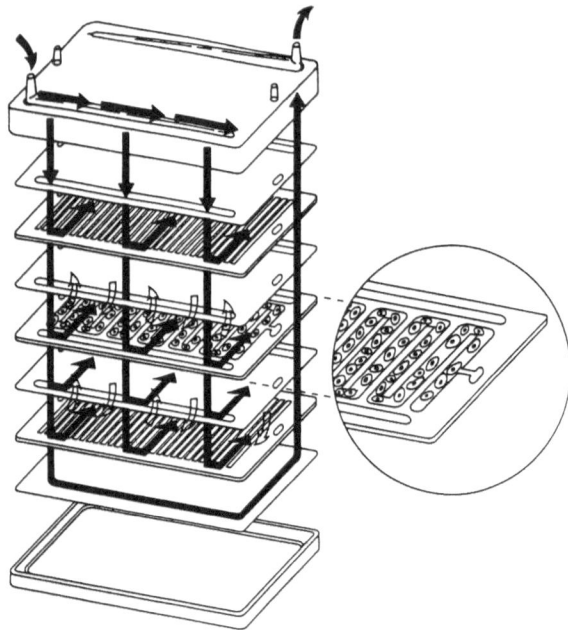

**Fig. 10.** Multiple layer flat sheet entrapped-cell bioreactor showing the zig-zag cell-containing channels alternating between medium flow channels (from Rainen [45])

studied a novel intercalated spirally wound hollow fiber bioreactor for mammalian cell culture which uses alternating sets of dead-ended fibers for medium addition and removal by forced convection [48].

In another application utilizing convective flow, Ozturk et al. [49] used two types of membranes in a transtubular bioreactor: silicone tubes for oxygenation and microporous polytetrafluoroethylene (PTFE) tubing for medium addition and removal. Two sets of PTFE tubing were used so that a variable portion of the medium could be forced through the cell suspension, achieving perfusion, by adjusting the pressure difference from one set of tubes to the other.

Although membranes have been placed inside suspension cultures for many years [50], a new development has been reported by Beyeler et al. [21] in which a porous stainless steel rotating spin filter is placed inside a draft tube together with an impeller-driven fan (Fig. 11). The fan drives the cell culture fluid up the draft tube along the surface of the spin filter, over the top of and down the outside of the draft tube, and back to the base of the draft tube. Fluid is sparged with gases both inside and outside the rotating spin filter. The gap between the spin filter and the draft tube is fairly large, on the order of the diameter of the spin filter itself, and hence Taylor vortices are probably not induced in such a system. Advantages of this bioreactor are a decoupling of the cell dilution rate from the medium dilution rate, excellent convective mixing, and low fouling due to axial and rotating flow across the sieve.

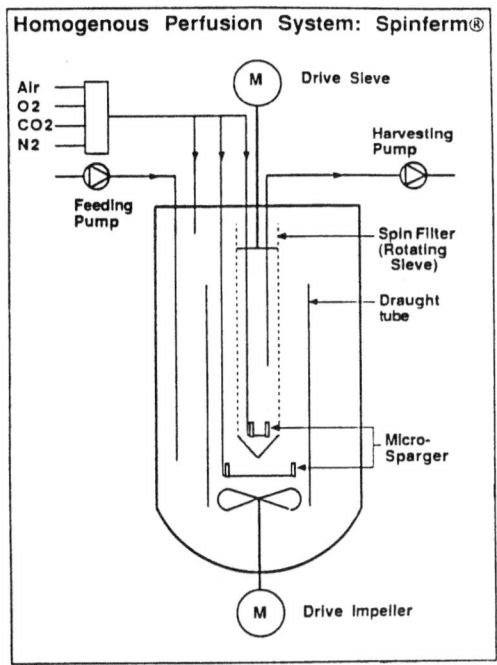

Fig. 11. Porous stainless steel rotating spin filter inside draft tube with impeller-driven fan (from Beyeler et al. [21])

Another new exciting membrane reactor is the Märkl et al. [51] tubular membrane filter system in which a dialysis membrane tube is inserted inside a strong sterilizable outer jacket. The suspension culture is maintained with an impeller either inside or outside the membrane tube and the product can be isolated following diffusion through the dialysis membrane (Fig. 12). If the cells are placed inside the tubular dialysis membrane, the medium can be fed to the external annular region at a dilution rate independent of the cellular growth rate. This flexibility allows for continuous supply of medium to and removal of waste products from the reactor, which should result in increased cell concentrations and growth rates. Another advantage of this new reactor design compared to conventional transparent glass bioreactors is that it can be safely sterilized in situ.

In an attempt to take advantage of the properties of nonpolymeric membranes, Kornfield et al. [52] suggested and tested a novel reactor concept which utilizes both ceramic and hydrophobic membranes for facilitated gas-liquid mass transfer. The monolithic (ceramic) reactor consists of two sets of channels, running orthogonally in alternating layers, which allow medium flow in one direction and gas flow in the other direction. A hydrophobic gas permeable membrane was used to separate the alternating layers to prevent mixing of the gas and liquid streams. The cells are immobilized on the porous ceramic support in the medium channels. High oxygen transfer rates with minimal power input

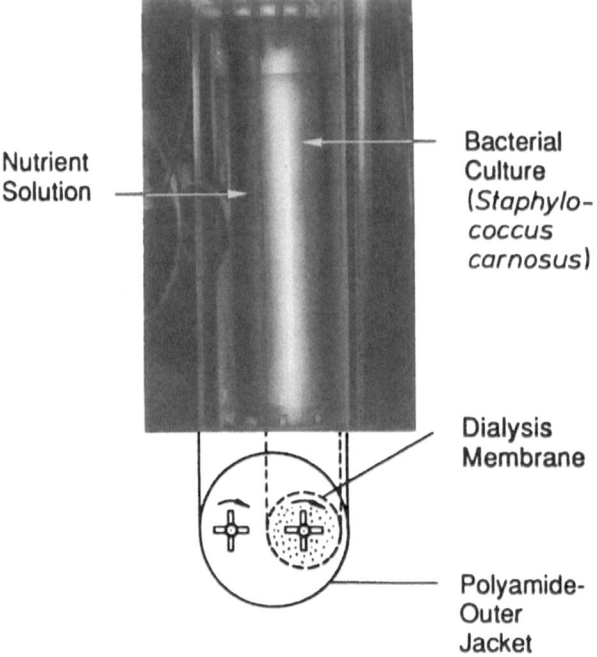

Nutrient Solution

Bacterial Culture (*Staphylo- coccus carnosus*)

Dialysis Membrane

Polyamide- Outer Jacket

**Fig. 12.** Tubular membrane filter system with dialysis membrane tube inserted inside a strong sterilizable outer jacket (from Märkl et al. [51])

was achieved. Ceramic membranes have already been successfully commercialized as whole cell reactors (e.g. Opticore, Charles River Biotechnical Services, Wilmington, MA)

In spite of the advantages of using membrane bioreactors for culturing cells and generating product, these systems do have their limitations [36]. Generally speaking, monitoring and control of perfusion cultures is more complex than for batch cultures. The high cell densities found in most membrane bioreactor configurations can result in diffusion limitations, lowering cell viability and volumetric productivity. One advantage of membrane bioreactors is the capability, in many cases, of using convective flow to overcome these mass transfer limitations. This has been studied theoretically by Schonberg and Belfort [53] and experimentally by Heath et al. [54, 55]. Overall mass transfer and reaction have been modeled for membrane bioreactors with the intent of improving design, operation and scaleup of such systems [33, 56–63]. These modeling efforts will be improved as we develop a better understanding of cellular metabolism and of the effects of cell density on cell growth and productivity. For example, it has recently been reported for a mouse-mouse hybridoma line (ATCC HB-124) that monoclonal antibodies produced by these cells inhibit their own secretion rate [64]. Clearly, removing them from the culture fluid should enhance antibody productivity. Membrane bioreactors with perfusion could be used, to good effect, in this case.

# 3 Membrane-Immobilized Enzyme Reactors

Because of their high packing density (large surface area per unit volume of reactor space), hollow fiber membrane systems are frequently used to retain enzymes for bioconversions [65–76]. Membranes may be used in two ways to entrap the biological catalyst while the substrates and products freely diffuse through the membrane: as a barrier for retaining the soluble enzyme or as a high surface area per unit volume matrix on which the enzyme is immobilized, the combination of which is sometimes called a "reactive membrane" [77].

## 3.1 Soluble Enzyme Membrane Systems

The enzyme in solution may actually consist of the enzyme itself or a whole cell, rendered partially porous while retaining the desired enzyme. The advantages of immobilizing soluble enzymes by a membrane barrier (e.g. in the shell space of a hollow fiber module) for catalytic conversions include high retention of enzyme activity, high rates of mass transfer, ease of enzyme replacement, simultaneous reaction and, in many cases, product isolation.

Many biosynthetic reactions such as those involving energy or group transfer, redox reactions or covalent bond synthesis require cofactors or coenzymes such as ATP, NAD(H), or NADP(H) in addition to the primary enzyme [37]. To render the biosynthesis economically feasible, these expensive cofactors must be recycled and regenerated for reuse. Membrane bioreactors have recently been used for the retention and recycle of these expensive and often unstable coenzymes in continuous operation [78–83].

Traditionally, the two most popular approaches involve (a) binding of the coenzyme to soluble macromolecules (such as PEG) which are retained by the membrane along with the enzymes, or (b) membrane retention of the enzymes without immobilization of the coenzyme. Using the latter process, Ishikawa et al. [78, 79] conducted both theoretical and experimental investigations of glucose 6-phosphate production with simultaneous ATP regeneration by enzymes in an ultrafiltration hollow fiber reactor. The enzymes for substrate catalysis and ATP regeneration were retained by the membrane in the reactor shell space. Substrate and ATP (low concentration) were continuously supplied through the fiber lumens. Substrate, product, ATP, and ADP diffused freely through the membrane. While ATP and ADP were not actually immobilized in the reactor, retention in the shell space was high because as soon as ATP was released as ADP from the primary enzyme it bound to the regenerative enzyme to form ATP which would bind again to a primary enzyme and the cycle repeated, i.e. the ATP and ADP were "dynamically recycled." Use of the hollow fiber enzyme retention process demonstrated improved space-time yield, recycle rate, recycle ratio, and substrate conversion compared to results obtained from CSTR-membrane reactors.

A newly developed charged membrane bioreactor has also been used for recycling coenzymes [80]. Free nicotinamide coenzyme was effectively retained by electrostatic repulsion between the coenzyme and the anionic surface of the membrane. Substrate and product molecules pass easily through the membrane while the enzyme and much of the coenzyme are retained. This system has been shown to be effective in the multi-enzyme production of sorbitol from glucose [80, 83] and in the production of gluconic acid and mannitol from glucose and fructose [81]. The charged membrane bioreactor has also been analyzed theoretically [84].

Microcapsules, or artificial cells as they are sometimes called, have also been used for retaining enzymes and their cofactors [85–89]. Gu and Chang immobilized a multi-enzyme system and dextran-NAD$^+$ for converting ammonia or urea into the essential amino acids L-leucine, L-valine, and L-isoleucine [89] in semipermeable microcapsules made from nylon-polyethylenimine. This was reported as far back as 1977 for generation of non-essential amino acids [85]. However, the more recent study has reported conversion of substrates into essential amino acids [89]. In this case, the dextran-NADH was continuously regenerated within the capsule.

In an enzymatic conversion not involving cofactors, Jones et al. [90] reversibly immobilized lactose in the spongy layers of anisotropic membranes in a hollow fiber reactor by backflushing the enzyme solution from the shell space to the lumen. The enzyme was retained in the spongy layer by the membrane skin while the products, unreacted substrates and solvent passed through the skin into the fiber lumen and flowed out of the reactor. The enzymes were reversibly immobilized and could easily be replaced. Good conversions were achieved without any loss in enzyme activity or stability. Jones et al. also developed a theoretical model of the system, which agreed well with the experimental data [90].

## 3.2 Reactive Membranes

Most entrapped membrane bioreactors rely on diffusion of substrate and/or product over relatively long distances (on the order of tens to hundreds of micrometers). Clearly, any reduction of this diffusion distance by using methods such as convective flow or mixing could substantially enhance bioreactor performance and result in significantly reduced membrane requirements. Microfiltration membranes, serving as a support for enzyme attachment, are referred to as "reactive membranes" and can be operated in either diffusive or convective transport modes [77]. Although problems of loss in activity upon immobilization of the enzyme to the membrane surface and of enzyme orientation or steric hindrance have been cited as drawbacks, the capabilities of running the processes continuously, of reusing the enzyme, and of utilizing convective flow are very attractive features. As methods of immobilization improve, the disadvantages of the process are expected to diminish. In general, overall process efficiency is improved by combining the reaction and separation steps, a result achieved when using membranes.

Schnabel and others have developed an enzyme reactor using porous glass hollow fibers (Bioran, Schott, Mainz) which has several advantageous qualities [7]. Amyloglucosidase was immobilized with glutaraldehyde onto an $NH_2$-modified glass surface and was stable for more than 80 days. The high porosity of the glass membrane yields a large surface area for immobilization of enzymes ($3-30$ m$^2$ cm$^{-3}$ reactor volume). The large pore sizes allow for immobilization of high molecular weight enzymes.

Nakajima et al. have developed a forced-flow (i.e., convective) membrane enzyme reactor in which the enzyme is immobilized on porous ceramic membranes [91]. By attaching the enzymes to the porous membrane surface, not only was mass transfer improved by utilizing convection rather than diffusion, but the convection was not limited by the significant pressure drops found in enzyme-immobilized bead-filled column reactors. A 10-fold higher productivity was observed in their system as compared to a conventional column reactor in which the enzyme was immobilized on beads.

As noted by Michaels [14], reactive membranes also have the potential to facilitate sequential catalytic reactions. Control of the flow rate through the membranes allows for both fast reaction times (i.e. not limited by diffusion) and short residence times, minimizing diversion of product into unwanted side reactions. By forcing the solution through a series of reactive membranes, a sequential reaction can be conducted with little back conversion due to the dominance of convective fluxes over diffusive ones.

Matsuura and coworkers [92] have immobilized yeast cells between an ultrafiltration (UF) membrane and a reverse osmosis (RO) membrane. Glucose substrate is passed convectively through the UF membrane, while the product (ethanol) and $CO_2$ pass from the cell layer through the RO membrane. Of course, polarization and fouling effects at both the UF and RO membranes could limit the reactor's long-term operation.

## 3.3 Aqueous/Organic Systems

For many reactions, substrates are often very soluble in organic solvents but sparingly soluble in aqueous solution. Hydrolytic enzymes such as lipases are known to be most active at the organic-aqueous interface. For reactions such as those in which dispersions are often used to increase the surface area between the organic and aqueous phases, membrane systems could be of advantage. The membrane can serve as a barrier between aqueous (enzyme-containing) and organic (substrate-containing) phases with excellent mass transfer and little if any inactivation of the enzyme.

An example of an aqueous/organic reactor, reported by Matson and Quinn in 1986 [77], is a laminated membrane system consisting of a non-catalytic layer (membrane-supported organic solvent) which is permeable to the substrate but not to the product, and an enzyme-containing layer which is permeable to both the substrate and the product. The feed is supplied on the noncatalytic side. The substrate diffuses through the noncatalytic layer into the enzyme-containing layer where it reacts, and the product, which cannot diffuse back through the noncatalytic layer, migrates into the sweep stream on the downstream side of the enzyme-containing layer. Matson and Quinn [77] used this procedure for enzymatic resolution of a racemic amino acid derivative. Product concentration can be controlled by the sweep buffer flow rate and, if the reaction thermodynamics are very favorable, the product concentration in the permeant stream can be higher than the substrate concentration in the feed stream [77].

The problems associated with dispersions, such as the difficulty in maintaining stability and efficient recycling of the organic phase, can be obviated with membranes. An excellent example of the application of such a system is the racemic resolution of the $R$ and $S$ isomers of glycidyl butryate [93]. Ladner and Whitesides [94, 95] have reported the synthesis of $R$-glycidyl butyrate by subtractive enzymatic resolution of racemic glycidyl butyrate using porcine pancreatic lipase (PPL). $R$-Glycidyl butyrate and its derivatives are useful synthetic intermediates for the beta-adrenergic blocker class of cardiovascular drugs. Lopez et al. [93] have entrapped the PPL in a hollow fiber polyacrylonitrile (PAN) membrane allowing the organic solvent (racemic glycidyl butyrate) and the aqueous solution to flow on opposite sides of the membrane. The $S$-ester is hydrolyzed faster than the $R$-ester and is converted to its acid by the PPL in the membrane; it then diffuses into the aqueous stream and is removed. The product $S$-ester was collected at 67% conversion with an enantiomeric excess of 96.7%. Wu et al. [96] have recently analyzed the mass transfer for this diffusion-coupled reaction system and predicted optimal design parameters and methods of operation. Their model is currently being tested with the glycidyl butyrate system, but should be tested and modified for other systems as well. For the glycidyl butyrate system, however, they show that increasing the enzyme loading results in a decrease in enantiomeric excess for subtractive resolution with increasing conversion. This effect is predicted by their model.

van't Riet and coworkers [97, 98] have studied lipase-catalyzed esterification and hydrolysis reactions in two-phase membrane systems. A hydrophilic membrane (cellulose) was used for separation of the phases as well as for immobilization of the enzyme, which in free form collects at the lipid-water interface. The lipase was immobilized on the "lipid side" of the membrane. Water for reaction permeates through the membrane to the enzyme layer; the product (glycerol) diffuses back through the membrane in the water phase.

Another advantage of using membrane systems for enzyme conversions is that the product can be removed in reactions which suffer from enzyme inhibition or unfavorable thermodynamics [99]. Multimembrane extractive fermentation of ethanol from yeast, discussed in more detail in Sect. 5.3, is an example of a feedback inhibited reaction which is made more favorable by the use of membranes to selectively remove the product. The product is extracted preferentially from the aqueous medium through a membrane into an organic stream which possesses a favorable solubility or affinity interaction, thus reducing product inhibition.

# 4 Membrane Bioseparations: Established Processes

## 4.1 Removal of Suspended Matter

In downstream processing it is often necessary to separate the cells from the culture medium either for recovery of the broth or for recovery of the cells. The recovered cell mass may be desired ipso facto or for the intracellular heterologous proteins which will subsequently be isolated. In all three cases, microfiltration is competitive with such processes as sedimentation, rotary drum filtration and decanting (Fig. 6). Usually the membrane process is external to the reaction tank and the culture broth is fed to the membrane system and recycled to the reactor after concentration. The permeate, usually containing the product, is then collected in a reservoir as shown in Fig. 13. Even with effective fluid mechanics, concentration polarization and/or membrane fouling occurs and results in a precipitous drop in the permeation flux from the pure water flux. Nagata et al. [100] observed an irreversible flux decline in the microfiltration of bacterial cell broth. In filtering the broth without cells in a 0.8 μm pore stainless steel microfiltration membrane, the permeation flux was found to drop from a value of about $2.7 \text{ cm s}^{-1}$ to less than $0.6 \text{ cm s}^{-1}$ during the concentration process (Fig. 14). Upon dilution of the broth, the permeation flux was not recoverable and remained at a value of about $0.75 \text{ cm s}^{-1}$ indicating the occurrence of serious permanent fouling. In order to explain the observed behaviour, Nagata et al. [100] identified the cause of the fouling and proposed a new fouling model called the Solids Flux Model. Apparently, after sterilization of the medium, a precipitate of $MgNH_4PO_4$ was deposited on and in the pores of the membrane. For this model, the log of the permeation flux is plotted against the solids concentration, as shown in Fig. 15a. Multiplying the flux by the solids concentration and plotting this product versus the solids concentration one obtains a direct measure of the amount of solids being deposited on the membrane (Fig. 15b). Once again, dilution of the feed as shown with the same data as in Fig. 14 did not result in complete recovery of the flux but instead the flux remained at a value of about $0.75 \text{ cm s}^{-1}$.

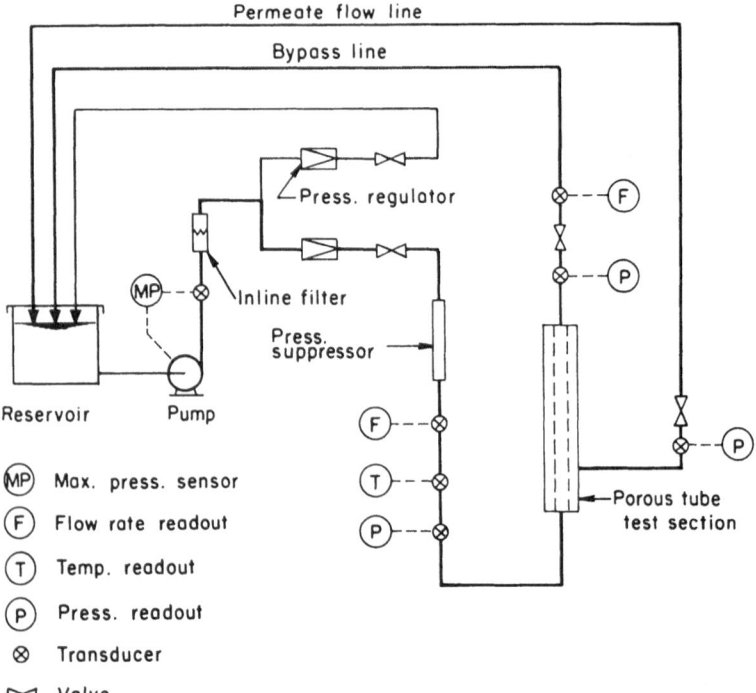

**Fig. 13.** Flowsheet of microfiltration test system showing the reservoir or bioreactor followed by the membrane unit (or porous tube section) (from Nagata et al. [100])

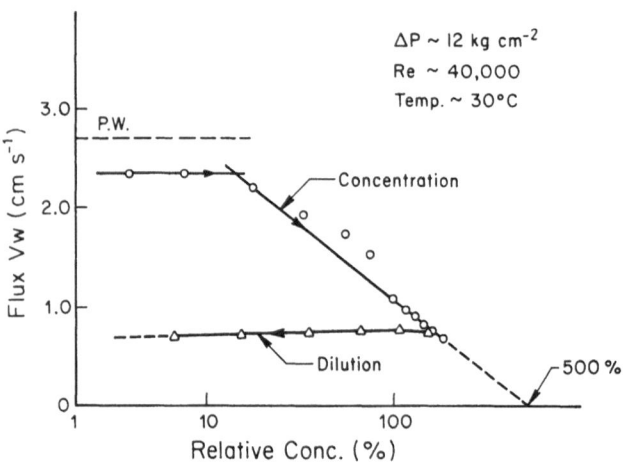

**Fig. 14.** Permeation velocity (wall flux) versus relative concentration of buffer solution during concentration and during dilution with pure water as feed (from Nagata et al. [100])

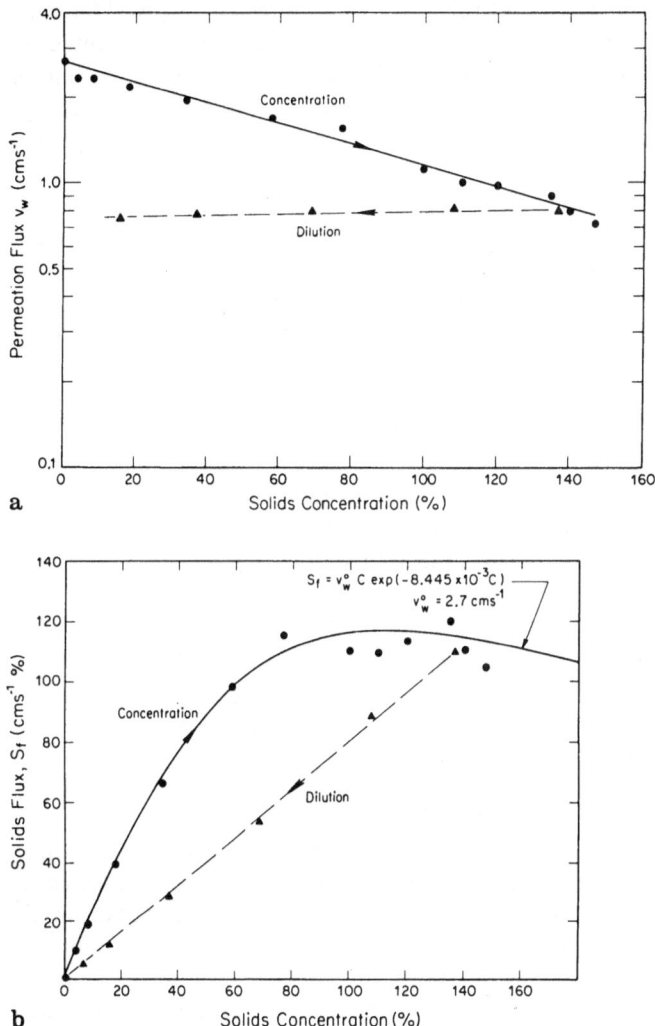

**Fig. 15a, b.** **a)** Permeation velocity (wall flux) vs relative concentration of buffer solution according to the solids flux model during concentration polarization with water as feed during dilution (from Nagata et al. [100]). **b)** Solids flux versus solids concentration according to the solids flux model (from Nagata et al. [100])

Cell recovery in membrane processes can be accomplished in hollow fiber [101, 102], plate and frame [101–103] and rotating devices [26]. Crossflow membrane filtration has been used extensively in cell harvest of bacterial cultures (see Hanisch [104]) however membrane filtration for recovery of animal cells has been somewhat limited due to cell damage and pore plugging by deformed cells and cell fragments. Extensive experimentation with a variety of membrane modules (flat plate and hollow fiber), cell lines, tangential flow rates (laminar shear), and transmembrane pressures by Maiorella et al. [102] have led to the conclusion that laminar shear is the primary mechanism of cell damage in crossflow filtration of animal cells. When cell recovery is the goal (or one of them) they found that cell damage could be avoided by limiting the average shear rate to less than 3000 s$^{-1}$. The transmembrane pressure must also be low enough to avoid plugging of the membrane pores by deformed cells.

van Reis et al. [105] have used microporous tangential flow filtration in hollow fiber membranes to isolate proteins from mammalian cells with cell recycle in an industrial scale process. In comparing traditional methods of protein isolation they found that conventional centrifugation was fast but the concomitant shear stresses resulted in cell rupture; low shear centrifuges were too slow. Flocculation and sedimentation or centrifugation could increase the speed but requires addition of an impurity which must be removed in a subsequent step and thus increases expenses. Liquid-liquid extraction was considered as a good possibility except that process parameters would need to be developed for each new product. Microporous tangential flow filtration was chosen because of its speed, high yield, potential for linear scale-up, cell containment, low operating costs, and the ability to use the same process for a variety of products [105]. Greater than 99% protein yields were obtained from 5000 l of culture while maintaining total cell number and viability. Control of the filtration rate and the transmembrane pressure was found to improve membrane performance by decreasing the extent of fouling [105].

Plate and frame crossflow membrane devices for the recovery of animal cells and their products have recently been commercialized (e.g. Amicon, Div. of W.R. Grace, Danvers, MA and Millipore Corp., Bedford, MA). The processors can handle from 10 to 50 liters of culture fluid at a flux of 90 l m$^{-2}$ h$^{-1}$ while keeping the average transmembrane pressure less than $0.7 \times 10^5$ Pa. According to Sheehan et al. [103], a suspension of human lymphoma cells was concentrated by a factor of 10 without affecting cell viability. They also found dramatic differences in performance with cell line, which they suggested may be a function of cell fragility which affects the quantity of cellular debris. For example, filtration of murine lymphoma cells at constant permeation rate demonstrated very little increase in transmembrane pressure while use of human lymphomas resulted in a significant pressure increase, presumably because of the greater amount of cell debris generated by the latter [103].

Rolghigo [106] and Kroner et al. [26] have tested bakers yeast with a rotating microfiltration filter using Taylor vortices and centrifugal pressure to depolarize and defoul the membrane. The former group used a positive displacement pump with the membrane on the outer stationary surface and an inner rotating surface without a membrane. With hybridoma cells and secreted monoclonal antibodies, they obtained steady permeation rates on the order of 500 and 1300 l m$^{-2}$ h$^{-1}$ for clarification with 0.8 and 3.0 μm pore size membranes. Tangential flow filtration is usually unable to attain such high permeation rates without severe flux declines. The main limitation of these rotating filters is the difficulty in scale-up.

## 4.2 Removal of Dissolved Components

Recovery of dissolved macromolecules such as valuable proteins from complex biological fluids, such as cell culture media, is a challenging problem. Membrane processes such as ultrafiltration have been successfully used in this regard (for a review of biocatalyst separation by ultrafiltration, see Ref. [107]). With secreted products (e.g. monoclonal antibodies and other proteins produced by mammalian cells), the first step in product isolation is separation of the product-containing medium from the cells. The next step is often concentration of the dilute solution, following by product isolation and purification steps. All of these steps can utilize membrane filtration. However, for product isolation, membrane fractionation can only safely separate molecules which differ by a factor of approximately ten in molecular weight because of the variability in pore size of most polymeric membranes and because of the variability in protein shape and chemical interactions.

For products which are not secreted into the medium (i.e. from many yeast and bacteria), it is necessary to recover the desired protein in the cell lysate after the cells have been disrupted. Microfiltration has been used to retain cell debris allowing the proteins and microsolutes to permeate the membrane. This permeate solution is then fed to an ultrafilter in which the microsolutes such as salts and sugars, etc. are allowed to permeate the membrane. The main limitation of this latter process is the increased viscosity of the concentrated solutions, especially near the membrane-solution interface. This increase in viscosity severely limits the capability of the process. The increase in concentration at this interface (concentration polarization) also imparts additional osmotic pressure, thus reducing the net driving force through the membrane [108–114]. Methods to overcome this have been developed also using membrane processes, including diafiltration. In diafiltration, low molecular weight contaminants are flushed through the ultrafiltration membrane and can be removed continuously with a feed and bleed process. Thus a two-step process of diafiltration, wherein the microsolutes are flushed from the system, and ultrafiltration, wherein the protein solution is concentrated, is utilized. The same process of diafiltration can also be used for changing buffer or solvents in a particular system.

Compared to concentration polarization, membrane fouling and plugging of the microfiltration and ultrafiltration membranes can be an even more serious limitation. Adsorption of proteins and other molecules on membrane surfaces affects both solute rejection and permeate flux rate. Particular cases of membrane fouling by the adsorption of proteins [115] and salts [100, 116] have been studied but more research is needed. With the development of asymmetric ultrafiltration membranes, especially the widely used polysulfone materials, higher water fluxes are attainable. However, because of the high potential for proteins to adsorb on hydrophobic polysulfone, fouling often reaches unacceptable levels. The most widely used microfiltration membranes are also made of highly hydrophobic polymers such as polyethylene and polypropylene. These also foul extensively and can adsorb proteins from solutions. However, because of their fairly large pore size, the effect of protein adsorption in the pores is relatively minor on their performance unless the pores become signficantly blocked and constricted. For this reason, serious attempts are currently being made to hydrolyze or convert the trunk polymers from hydrophobic to hydrophilic materials [117]. Either new hydrophilic materials are being used or the surfaces of standard polymers are being modified by attaching hydrophilic groups (e.g. hydroxyl, amide, epoxide).

In some cases, however, polymeric materials cannot withstand the pressures needed for the separation process. Chan et al. [118] used a stainless steel tubular membrane for isolation of a heterologous enzyme overproduced by E. coli. The stainless steel membrane was needed to withstand the high axial flow rates (mean shear stresses on the order of 12 to 47 Pa) required to disrupt chemically-weakened bacterial cell walls and to avoid fouling the membrane. The enzyme yield was close to 100% and the filtrate turbidity was reduced to only 4% that of the retentate. The permeate flux increased from 2000 to 9000 kg m$^{-2}$ h$^{-1}$ with increasing axial Reynolds numbers from 10,000 to 60,000.

In another recombinant DNA process, tumor necrosis factor (TNF), which was expressed in E. coli as a soluble protein, was isolated using a hydrophilic (cellulosic) membrane separation followed by a hydrophobic one (PTFE [119]). The process of lysate clarification by the cellulosic membrane followed by PTFE diafiltration is reportedly used as the basis for a large scale process in producing TNF for clinical trials (Cetus, Emeryville, CA).

Antifoams are often used in bioreactors to reduce foaming at gas-liquid interfaces. These antifoams often cause havoc in downstream processing by substantially reducing the flux through membranes. The reason for this is that antifoams are surface active and, hence, find themselves at the interface of the membrane surface and the solution, effectively fouling or blocking the pores. McGregor et al. [120] have demonstrated the effects of various commercial antifoams on ultrafiltration using three different molecular weight cutoffs for polysulfone membranes and found large variation in the degree of flux reduction. Pure silicone antifoams were found to have the least effect on flux for polysulfone membranes.

Clearly, the search for methods of reducing the interactions between proteins or other surface active molecules and the membrane is an area of prime importance in bioseparations research.

# 5 Membrane Bioseparations: Emerging Processes

Newer membrane separation processes utilize the microstructure and surface chemistry of membranes rather than the more obvious and traditionally used properties such as sieving [99]. Very high selectivity and relatively good flux have been obtained in immobilized ligand-adsorptive (affinity) membranes or by membranes using an extractant with coupled active transport. Aqueous two-phase extraction plus ultrafiltration has been used for recycling the phases and recovering the product in the membrane unit. Bipolar electrodialysis membranes have also recently been studied and developed to convert water soluble salts to their corresponding acids and bases [121].

## 5.1 Membrane Chromatography

Membranes are currently being used for chromatographic supports as an alternative to the conventional bead-filled column configuration because of their large surface area per unit volume and their greater range of flow rates. The pressure drop, and thus the flow rate, over the membrane module is limited by fiber strength, not by gel bead compaction as in conventional chromatography, thus allowing more control over the rate of convective flow.

Milby et al. have developed an ion exchange process using hollow fiber membranes which were modified (to minimize nonspecific adsorption of proteins) and functionalized to introduce charged groups for adsorption [122]. The process combined high adsorption partition coefficients with a large bed volume of minimal depth, the preferred geometry for ion exchange adsorption processes. The protein separation was achieved by radial flow through the hollow fiber membranes. By adjusting the pH and ionic strength of the feed solution, selectivity could be manipulated. After adsorption and rinsing, the protein was eluted, as usual, by changing the ionic strength or pH of the buffer. The process was accomplished in minutes and boasted high throughputs and protein recoveries of greater than 90%.

Proteins have also been separated by simultaneous ultrafiltration and adsorption (by mechanisms of ion exchange or affinity) onto gel particles placed in the shell space of a hollow fiber membrane module by Molinari et al. [123]. During the adsorption step, crude solutions were passed through the lumen and recirculated. Because of a pressure differential established across the membrane walls at one of the shell space exits, part of the recirculating fluid was forced through the membrane into the extracapillary region, which contained the adsorbent gel. In addition to the forced convective flux, proteins also entered the shell space by diffusion and Starling-type convective fluid flow. The adsorption process continued until the concentration of protein in the recirculating fluid decreased to a limiting value. To elute the protein, high ionic strength buffer was pumped through the system. An advantage of the process is that is does not require pretreatment of the feed and can thus be used on culture broth directly from the reactor. According to the authors, the process should have the same capacity as a chromatographic column of the same size without the inherent pressure drop limitations. The process, which was tested with an ion exchange matrix (DEAE-Sephadex) and an affinity matrix (an affinity gel Sepharose), reportedly achieved separations comparable to those achieved with column chromatography [123].

Hollow fibers without radial convective flow can also be used for liquid chromatography. Ding et al. [124, 125] have used hollow fiber liquid chromatography to separate solutes such as flavors and proteins because of their lower pressure drop as compared to conventional column chromatography. The pressure drop for a packed column with 30% void volume is more than 80 times that in a fiber bed [124]. Modules containing 120 to 27,000 parallel fibers of 100 μm diameter were able to separate myoglobin and cytochrome-c in the mobile aqueous phase using an octane solution of reversed micelles supported by a polymer membrane (microporous polypropylene) as the stationary phase. Excellent reproducibility was observed for the separation of myoglobin and cytochrome-c. More experiments are needed to determine the feasibility and capacity of the process for other separations. If the process is successful, it could prove to be a significant advance for preparative liquid chromatography since scale-up simply consists of adding more fibers or modules.

## 5.2 Affinity Membranes

Affinity separations, because of their high specificity, are particularly suited to high resolution purification of proteins from dilute and complex solutions such as cell culture media. Purification of proteins by affinity techniques has become increasingly popular in recent years because of improvements in the process such as diminished ligand leakage, increased availability of ligands, better methods of immobilization, and reduced ligand cost. However, affinity separations are traditionally carried out using immunosorbent beads in packed columns and thus suffer from low throughput because of mass transfer and pressure drop limitations. Volumetric flow through the column is limited because of the time required for diffusion of protein into the immunosorbent beads and because of bead compaction at high pressure drops. High flow rates result in inefficient use of the beads, lowering the effective capacity of the column. Use of smaller beads decreases the distance and time required for diffusion but increases the pressure drop across the column at constant flowrate.

By attaching the ligands to the pore wall surfaces of membranes, the high specificity of affinity interactions can be combined with the excellent transport properties of thin film membranes resulting in high separation efficiencies with high throughput. Molecules can be isolated from others of similar size due to affinity interactions, which cannot be accomplished using membrane filtration alone. The fast convective flow rates through membranes avoid the diffusive mass transfer limitations of traditional column chromatography, facilitating rapid large scale separations. The use of crossflow operation, and the ability to handle suspensions, could reduce the number of purification steps and improve the prospects for scale-up to industrial quantities. The interaction between the ligand and the ligate can be general, such as lectin:carbohydrate $(K_a = 10^4 – 10^6 \text{ M})$ or specific, as with antibody:antigen $(K_a = 10^6 – 10^{12} \text{ M})$ or enzyme:substrate $(K_a = 10^4 – 10^6 \text{ M})$ [126]. The higher the association constant,

$K_a$, the higher the binding strength between the two molecules. When the ligand sites are saturated, the ligate can be eluted by increasing the eluant salt concentration, decreasing the pH or eluting with an excess of a competing ligand. The membrane can then be washed and reused. Affinity membrane systems have recently been commercialized for bioseparations [127].

Three main advantages have recently been recognized in using membranes as active adsorption matrices. These are the very high internal surface area of microporous membranes, the ability of membranes to fractionate and separate suspensions, and the advantage of using convective flow rather than diffusive flow as is often the case in affinity adsorption processes with beads in columns. Membranes are also used both before and after chromatographic processes in order to remove suspended matter and to concentrate the relatively dilute product stream, respectively.

Essentially, the membrane can be considered as a flat porous bed with an extremely short fluid path. Ligands are covalently attached to the membrane pore walls which have pore diameters of 0.5 to 1.0 μm, large enough to permit convective flow and thus alleviate diffusion limitations. Scale-up is easily achieved when using hollow fiber membranes by increasing the number and/or length of fibers with constant bed height (membrane thickness). By linear scaling of the flow rate with respect to matrix volume, the fluid residence times, the breakthrough behavior, the transmembrane pressure and the volumetric productivity remain constant as the device size is increased. In an application of the process, Matson indicates that isolation of 80 mg (97% yield) of murine monoclonal antibody from clarified cell culture supernatant can be achieved in 15 minutes using a Protein A affinity membrane module [127]. An agarose affinity column of similar size would yield a productivity approximately two orders of magnitude lower. Zale et al. [127] claims the process is up to 100 times faster than conventional chromatography. Because volumetric throughput is no longer a limitation, the process can now be used earlier in the recovery sequence, eliminating concentration and partial purification steps.

Fixing charged groups to a membrane surface has been popular since the development of permselective ion exchange membranes in the early 1950s [128]. Cationic exchange microporous membranes have recently been commercialized and used to capture large molecules such as monoclonal antibodies from cell culture fluid. Biospecific adsorption using ligands such as antibodies or metal chelates ($Cu^{2+}$, $Ni^{2+}$, $Zn^{2+}$) are currently being investigated to develop a whole cast of bioreactive membranes. Iwata et al. [129] and Serafica and Belfort [130] have developed an immobilized metal affinity hollow fiber membrane to isolate proteins. Using porous glass substrates (Bioran, Schott Glaswerk, Mainz) with modified surface chemistries (i.e. with epoxy or amine surface groups), Serafica and Belfort [130] have attached iminodiacetate (IDA) – metal (Cu, Ni, Zn) complexes to the glass surface. These materials were then used to chelate proteins that had histidines, cysteines and/or tryptophans on their outer surface from solution. They obtained, as expected, distinctly different adsorption isotherms for lysozyme, RNase A, cytochrome C and alpha-chymotrypsinogen A.

After optimization, these immobilized metal affinity (IMA) chelate membranes could be widely useful for fractionating proteins from complex media applied in microbial and animal cell cultures.

Borrowing from successful column processes, an affinity membrane consisting of Cibacron blue dye bound to a supported synthetic copolymer membrane has been used for protein purification (Sartorius, Göttingen, Germany). When compared to conventional affinity chromatography with Blue Sepharose, the Sartobind Blue membrane, used under crossflow conditions, yielded better performance in isolating malate dehydrogenase from a cell homogenate of *E. coli* [131]. Using the membrane, a purification factor of 200 and a yield of greater than 90% were achieved. In batch adsorption, the capacities of the membrane and Blue Sepharose were similar (per milligram of dry material) but the enzyme recovery was significantly greater for the membrane during crossflow filtration. The membranes were used for over 40 cycles successfully with the initial capacity restored completely after regeneration with NaOH. The initial capacity was not regained for the Blue Sepharose gel following regeneration. Higher flow rates (shorter residence times) had little effect on the binding capacity of the Sartobind Blue membrane as compared to the Blue Sepharose gel which suffered from diffusion limitations. The Sartobind Blue membrane also resulted in significantly higher recoveries of enzyme from cell homogenates when used in the crossflow filtration mode than the Blue Sepharose gel.

Another method of protein separation using affinity membranes has been termed "affinity-mediated membrane transport" in which a "switch" monoclonal antibody is used as a highly selective protein carrier in facilitated transport through a liquid membrane [132]. Switch monoclonal antibodies can experience a significant change in affinity for its antigen in response to a small change in the environment (temperature, pH, ionic strength) [133]. The physicochemical environment of the supported liquid membrane is controlled to yield a high antibody-antigen binding affinity on the upstream side of the membrane and a low binding affinity on the downstream side leading to complexation on the upstream side, diffusion of the complex across the membrane, and then decomplexation on the downstream side. The carrier molecule, which remains inside the membrane, then returns by diffusion to the upstream side to pick up another antigen molecule and repeat the shuttle cycle. Local pH is used to control the affinity of the antibody for the antigen. The Donnan potential was found to enhance the flux rate over the short range while an external electric field could be used to improve the flux rate for larger scale processes. A monoclonal antibody and its antigen, human growth hormone, were used to test the process [132].

An alternative combination of affinity and membrane processes is to use the membrane as a barrier which retains the ligand on one side rather than as a binding support for the ligand. Herak and Merrill [134, 135] have theoretically and experimentally investigated batch operation of affinity crossflow filtration. Their system consists of a stirred tank in which the "affinity escort" (other names for the bound ligand complex used in the literature include macromolecular ligand [136, 137] and macroligand [138]) binds the desired biomolecule, then the nonbound solutes are washed away, and finally the biomolecule is eluted from the affinity escort. The fluid is then pumped to the filtration unit where the biomolecule passes through the membrane in the filtrate, and the affinity escort is recycled back to the stirred tank. The setup appears to be a modification of the continuous affinity recycle system studied by Pungor et al. [139] a few years earlier (Fig. 16). The affinity escort in this case is Cibacron Blue-agarose and the biological molecule is human serum albumin (HSA) [134]. Later work by Herak and Merrill [135] also included studies of Cibacron Blue and lysozyme, protein A and IgG, and concanavalin A and horseradish peroxidase. In a similar manner, Mattiasson and Ramstorp [140] have used affinity crossflow filtration (ACFF) to isolate highly purified concanavalin A from a crude extract of Jack beans using heat-killed yeast cells as the affinity macroligand or escort, with a yield of 70%. Luong et al. [138] created a high molecular weight polymer with an affinity for trypsin. In batch ACFF, they isolated trypsin from chymotrypsin with a yield of 90% and a purity of 98%. The continuous process is still under study. Dextran [137, 141], silica nanoparticles [142], and starch granules [137] have also been used to form affinity escorts.

A variation of membrane affinity separations involves the use of membrane-encapsulated soluble ligand conjugates. Sakoda et al. [143] have recently reported on the encapsulation of soluble and insoluble ligands Blue Dextran

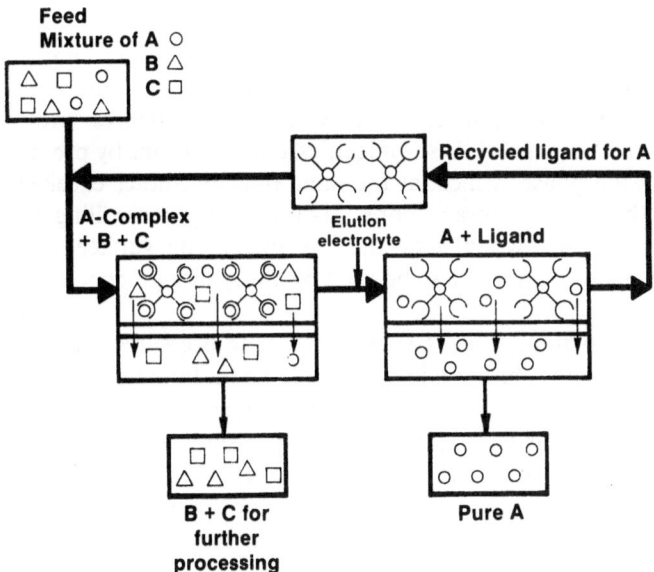

**Fig. 16.** Continuous affinity ligand membrane extraction showing the filtration unit where the non-bound microsolutes are washed out of the feed stream and the stripping unit where the desired biomolecule is released from its ligand. The biomolecule then permeates through the membrane leaving behind the ligand which is recycled for further complexation (after Mattiasson and Ling [137])

and Blue Sepharose, respectively, in calcium alginate membranes. By encapsulating the ligands with a hydrophilic membrane, nonspecific adsorption of macromolecules and other contaminants can be avoided. The use of soluble polymer-ligand conjugates rather than solid adsorbents in the capsules minimizes internal diffusion resistances as shown by the greater isolation of human serum albumin by soluble Blue Dextran over the insoluble Blue Sepharose. The membrane properties can be manipulated to retain the soluble ligands while allowing easy diffusion of the product molecule into the capsule.

Certain downstream processing problems lend themselves to the use of affinity membranes. Media containing low concentration of the desired protein or a high concentration of contaminating proteins (e.g. serum proteins) are two such examples. Processes which combine reaction and separation are also candidates for membrane affinity systems.

Affinity crossflow filtration, because of its ability to process unclarified and viscous media, has the potential to be a great success. The major holdup is approval of the ligands by regulatory agencies. Ligand design, such as with antigen-monoclonal antibody interactions is surely on the horizon for practical biotechnology separations. More work also needs to be done to reduce costs and to develop additional ligands. The reader is referred to the literature for additional details [144].

## 5.3 Membrane-Assisted Extraction

Conventional solvent extraction utilizing dispersion of one phase in another often results in some degree of phase entrainment and is limited by the small density difference and the interfacial tension between the two phases. Membrane-mediated extraction can avoid these difficulties and limitations by providing a barrier between the phases which, at the same time, promotes excellent interfacial mixing within the membrane pores [145, 146]. By controlling the volumetric flow rates of the two phases and the membrane surface area, significantly different phase volumes may be used resulting in high concentration factors.

The process is easy to implement. Using appropriate pressures for each phase, membranes act as phase barriers in immiscible liquid-liquid extraction. Achieving phase separation with a hydrophilic membrane requires that the pressure of the organic stream be greater than the pressure of the aqueous stream and that the pressure of the organic stream be less than the intrusion pressure into the membrane (estimated with the Young-Laplace equation). If the pressure of the organic stream is larger than the intrusion pressure, the organic phase will enter the hydrophilic pores. If the pressure of the aqueous stream is greater than the pressure of the organic stream, then the aqueous phase will ultrafilter through the membrane [147]. Both of these effects are undesirable.

Studies of differential protein extraction by Dahuron and Cussler [148] have shown that hollow fiber extractions are substantially faster than those conducted in conventional equipment. Hollow fibers have a very large surface area per unit volume (a) leading to a high overall mass transfer coefficient ($K_La$) even though $K_L$ itself may be rather low. Because the lumen and extracapillary space flow rates can be controlled independently, flooding and channeling are not important.

Prasad and Sirkar [149] used microporous hollow fibers and flat membranes to effect dispersion-free solvent extraction of a bioconversion-based pharmaceutical product using a pH swing procedure. The change in pH resulted in a change in the solute distribution coefficient. They indicated that the advantages of hollow fiber extraction include variation of individual flow rates without flooding, no need for density differences between phases, high interfacial area, ability to handle particulates, and avoidance of emulsions. The highest interfacial area will result from small diameter fibers however this also yields a significantly greater pressure drop, necessitating a compromise. It is also possible, depending on the pressures and surface tensions of the two phases, that the non-wetting phase will pass through the membrane and enter into the wetting phase, creating an emulsion. This should be controllable by judicious choice of solvents, membrane pore size, and operating pressures [149].

Basu and Sirkar [150] have used hydrophobic hollow fiber membranes to extract citric acid from an aqueous solution by reversible chemical complexation of the citric acid with tri-n-octyl amine in an organic solvent. The process has been modified to make it continuous by Sengupta et al. [151] using two sets of hollow fiber membranes acting as phase barriers between the immiscible liquid phases (Fig. 17). The feed solution enters one set of the hollow fiber membranes. The solute is extracted into a second phase in the extra-capillary space and is then reextracted into the strip solution in the second set of fibers. Relatively high extraction rates with a small solvent volume can be achieved if rapid solvent recirculation rates are used. To further expedite the extraction, the pH in the aqueous streams can be varied independently. The process can also handle particulates, e.g. the feed can come directly from a bioreactor. Sørenson and Callahan [152] have used a similar system for penicillin isolation and found that sparging of the stationary liquid phase increased the mass transfer coefficient, indicating the presence of transport resistances. In a similar process, Dordick et al. [153] have used enzyme-facilitated transport through a liquid membrane to selectively separate and purify organic acids.

While membranes have been incorporated into bioreactors for simultaneous production and isolation as an end in itself, they have also been used for reducing the effects of product feedback inhibition. Jeon and Lee [154] used membrane-assisted extraction to isolate butanol from the fed batch process of *Clostridium acetobutylicum*. The effects of product inhibition were reduced and a fourfold increase in productivity over straight batch operation was obtained.

Shuler et al. [155–157] have used a multimembrane bioreactor for extractive production of ethanol from *Saccharomyces cerevisia*. Flat membranes were used to separate the gas phase (for addition of $O_2$ and removal of $CO_2$), the cells, the

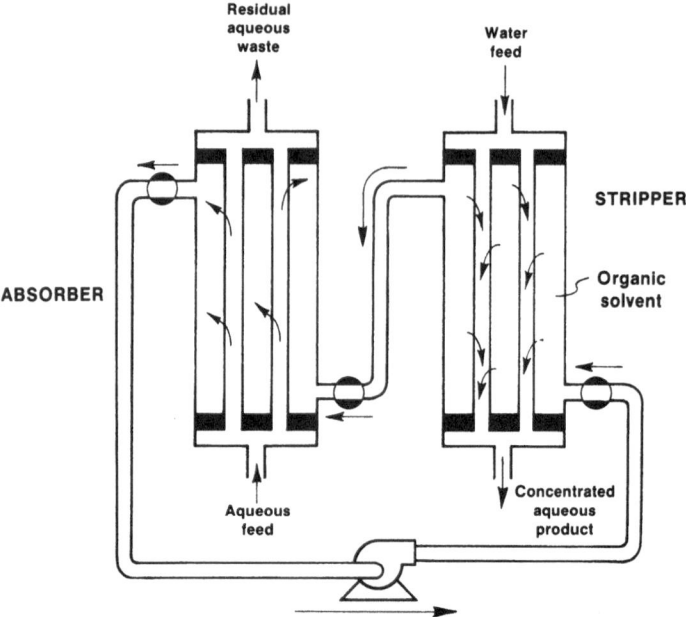

**Fig. 17.** Membrane solvent extraction showing the absorber in which the desired solute is extracted from the feed stream through the membrane into the organic extractant and the stripper in which the product is re-extracted out of the organic solvent into a concentrated aqueous stream (after Sengupta et al. [151])

aqueous medium, and the organic extractant. In one system, hydrophobic membranes were placed between the gas and cell layers and between the medium and extractant layers while a hydrophilic membrane separated the cells and medium. Nutrients from the medium were made available to the cells by both diffusive and pressure-driven flows. The rate and extent of growth were both enhanced [155]. The reactor performance was improved by implementing pressure-cycling in which the nutrient and extractant phases were convectively forced into and out of the cell layer [156]. Modifying the system by separating the reaction and extraction steps, Steinmeyer and Shuler [157] have maintained successful continuous operation for over 3000 h.

Other developments include liquid membrane emulsions in a capsule configuration [158–162] in which two miscible aqueous phases, the inner phase containing the enzyme and product and the outer phase containing the substrate, are separated by an organic phase. The organic phase forms a capsular liquid membrane through which the substrate must diffuse to react with the enzyme(s). The enzyme emulsions thus combine enzymatic reactions with selective transport through the organic phase, the rate of which can be controlled by substrate solubility in the organic phase or by facilitated transport of the substrate through the organic layer. Scheper et al. [158] used this process for continuous production of L-isomers of amino acids from a racemic mixture.

Quaternary ammonium salts were used as a carrier to transport the L-amino acid through the organic membrane phase.

## 5.4 Electrically-Driven Membrane Processes

Electrodialysis has been used for many years in desalting of cheese whey and of salt-precipitated proteins (e.g. ammonium sulfate). The advantage of using electrodialysis over gel permeation chromatography is its concentrating effect, leading to ease of salt reuse [163].

Electrodialysis can also be used for membrane modulated precipitation of proteins. In this process, the protein solution is allowed to flow through the concentrated duct of an electrodialysis unit in which ions such as ammonium sulfate are electrically driven into the stream at a controlled rate. As the protein solution progresses along the flow path, proteins will precipitate in a manner which is controlled by the rate of salt influx. Electrodialysis across ion-selective semipermeable membranes has also been used to force isoelectric precipitation of proteins from human plasma [164]. This protein fractionation technique was used to collect or remove trace proteins without affecting the physicochemical or functional properties of the proteins.

Amino acids have been separated by electrodialysis or by electrotransport through ion exchange membranes. Gavach et al. [165] have used this process to isolate alanine from other amino acids following acid hydrolysis of protein residues under different pH conditions with good results.

The use of electric fields has been proposed as a method for enhancing crossflow filtration by altering some of the effects of membrane fouling [166]. The electrical field is used to alleviate part of the membrane cake resistance due to electrophoretic motion of the charged particles (cells) away from the membrane, enhancing flux rates. For this technique to be effective the particles must have a significant net charge and the medium should be low in conductivity, which may be problematic in aqueous phase separations. Using a modified electrodialysis module, Brors and Kroner [166] demonstrated the utility of this technique using yeast by reducing the flux decline over time. The electrokinetic effect is strongly dependent on cell type, cell state, and culture fluid conditions. Despite obvious limitations, this process has potential and is a step in the right direction towards developing techniques to reduce the nemesis of membrane filtration, i.e. flux decline due to concentration polarization and fouling.

Electrical forces have also been used for dynamic control of protein transport across a hydrogel membrane by directly altering the solute flux and the membrane microstructure. Grimshaw et al. [167] found that protein transport could be significantly affected by electromechanical and swelling phenomena in polyelectrolyte membranes. They identified four distinct reversible mechanisms for controlling solute flux: electroosmotic and electrophoretic augmentation of solute flux through the membrane, electromechanical deformation (swelling) of the membrane, and electrostatic partitioning of charged solutes into the charged membrane. The process has the potential to effect selective, dynamic control of solute transport by alteration of the chemical structure of the membrane, the buffer composition, and the electrical gradient across the membrane.

## 5.5  Membranes and Precipitation

Membrane filtration has recently been demonstrated as an alternative to centrifugation for the collection of protein precipitates [168–171]. Precipitation of the protein prior to filtration has been shown to increase the membrane flux rates over that achieved with unprecipitated protein suspensions. Studies of ultrafiltration of soya protein precipitates in unstirred batch cells [169], and under crossflow conditions in hollow fibers on the laboratory [168] and the pilot [170] scales, and with microfiltration flat sheets [171] have been reported. Important parameters were found to include the precipitate particle size and the membrane rejection of soluble protein left in solution. Minimizing the amount of soluble protein resulted in greater flux rates, as did the use of suspensions with larger particle diameters. Both the inertial lift and shear-induced diffusion theories predict better performance with larger suspended solutes than with smaller solutes [29, 172]. The solution pH has also been shown to be very important for this process, particularly when it is near the isoelectric point of the proteins [171].

Recombinant DNA technology has resulted in a different type of "precipitate" that has also been recovered using membranes. Cloning and expression of foreign genes in E. coli often results in the formation of inclusion bodies which are insoluble aggregates of protein found within the cells. Inclusion bodies have been isolated from cell lysates primarily by centrifugation but Forman et al. [173] have studied crossflow filtration as a means of recovery. The presence of the retained inclusion bodies significantly affected the flux of soluble protein through the membrane such that an increase in the crossflow rate did not increase the soluble protein flux as predicted by concentration polarization theory. By avoiding high transmembrane pressures and the concomitant soluble protein retention, continuous concentration and washing of the inclusion bodies using crossflow diafiltration was found to be an attractive alternative to centrifugation.

The use of membranes during the actual precipitation process rather than simply as an alternative to centrifugation for collection of precipitates has been suggested by Michaels and Matson [99]. Membrane-modulated precipitation has been proposed to overcome the difficulties in controlling the concentration of the precipitating agent (electrolyte or solvent) during the formation of precipitates. Electrodialysis or dialysis can be used to control the transfer of solvent or electrolyte into the protein solution, effecting higher efficiency and greater selectivity.

## 5.6 Cryofiltration with Membranes

As a relatively new and untested idea, cryofiltration can be used in conjunction with membranes to achieve some separations. In this process, as shown in Fig. 18, an ice slurry is produced in an evaporative freezer which is then passed through a cascade of filters. Since the ice particles are relatively large, they do not pass through the membranes, while the non-ice-forming solvent does. The ice can then be passed to a melter to recover the ice-forming crystals in solution. Pure water is obtained from the melter while the product concentrate is obtained as the permeate through the filters [14].

# 6 Where to Now?

It should be clear from this review that membranes are of growing importance in both upstream and downstream processes in biotechnology. Many of the recent advances combine one or more of the attractive features of membranes with other reaction or separation methods resulting in an improved process. These incorporations are not only possible but preferred because of the numerous advantages accorded older, established processes when modified in some way by addition of a membrane process. The combination frequently increases process

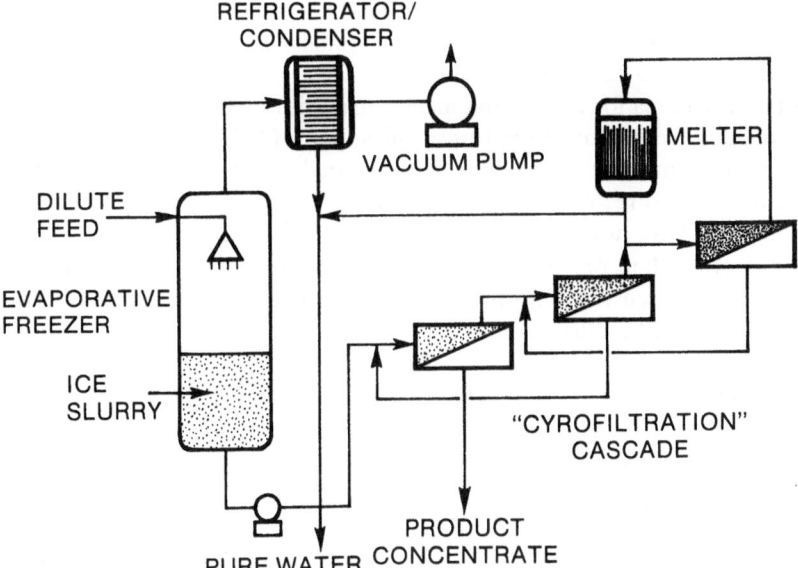

**Fig. 18.** Concentration by cryofiltration showing the recovery of ice crystals from the membrane cascade in the retentate and the non ice-forming product concentrate in the permeate (after Michaels [14])

flexibility by decoupling one or more variables from the rest, increasing the degrees of freedom.

It should also be apparent that particular membrane features are especially desirable for use in bioseparations. Narrow molecular weight cut-off membranes would result in fractionation of various protein solutions more accurately. Ultrathin active layers would result in increased permeation rates. Advances are continually being made in these areas. Examples of new membranes that meet these two criteria are S-layers, membranes from Langmuir Blodgett ordered layers, reverse micelles, electrochemically prepared ceramics and precipitated skins on ceramic-based membranes (not discussed in this review).

Another issue which needs further study is development of low-fouling potential surfaces. In order to understand the interaction of proteins with different polymeric and organic surfaces it is of crucial importance to measure the interactive forces between such systems. This is currently being pursued by Belfort, Pincet and students at Rensselaer.

In order to increase the productivity for catalytic or active membranes, it is desirable to obtain very high density catalytic or sorptive surfaces. In order to do this, it is our contention that oriented molecules such as F(ab) fragments offer some possibilities in this direction.

It is important that we learn how hydrodynamic instabilities such as Taylor vortices induced in the annulus of a rotating filter can help in the design of new modules. These new designs could be used specifically for biotechnology. One of the limitations of the Taylor-vortex annular filter is the difficulty in scale-up. In general, the gap size is restricted to be much less than the cylinder radii and, therefore, scale-up is most likely to occur in such systems by increasing the length. Unfortunately, axial rotation of such systems is mechanically difficult. The process may also become expensive.

One potential area for integrating membranes with other processes is in the recovery of valuable proteins from transgenic animals [174]. Here, complicated solutions such as milk or blood need to be treated using several coordinated processes in order to recover the desired proteins.

Although we have not discussed new and improved uses of membranes for applications other than reaction and separation in this paper, membranes are finding new uses in sensors, both in liquid and gas phases and as switchers for intelligent membranes [175].

*Acknowledgements*: Many ideas of A.S. Michaels and S.L. Matson are incorporated into this review. We especially acknowledge their work. We would also like to thank Robert Furneaux, Alcan; Robert Goldsmith, CeraMem; T. Alan Hatton, MIT; Helmut Ringsdorf, University of Mainz; Roland Schnabel, Schott; and Mary K. Tripidi, Kinetek for sending us their reprints, etc. In addition, we especially thank Steven L. Matson, Arete Tech., for sending us preprints, slides, and a copy of this class notes for the University College London's short course. Part of this paper was presented by GB at the European Biotechnology Conference, No. 5, Copenhagen, Denmark, July 1990.

# 7 References

1. Nagase Y, Takamura Y, Sugimoto K (1990) Improved alcohol permselectivity for pervaporation with modified poly(1-trimethylsilyl-1-propyne) membranes. Proceedings from the International Congress on Membranes and Membrane Processes (ICOM), August 1990, Chicago, vol 1, p 331
2. Sedath RH, Funk EW, Li NN (1990) Reduction of fouling in ultrafiltration membranes via surface fluorination. Proceedings ICOM, August 1990, Chicago, vol 1, p 106
3. Nagase et al. (1990) J Polym Sci, Part B. Polymer Physics 28:377
4. Le MS, Gollan KL (1989) J Memb Sci 40:231
5. Clark WM, Bausal A, Soutakke M, Ma YH (1991) J Memb Sci 55:21
6. Sara M, Sleytr UB (1987) J Memb Sci 33:27
7. Schnabel R, Langer P, Bayer E (1989) Application oriented treatment of inorganic membranes. 6th International Symposium on Synthetic Membranes in Science and Industrie, Sept. 4–8, 1989, Tübingen
8. Albrecht O, Laschewsky A, Ringsdorf H (1985) J Memb Sci 22:187
9. Goldsmith RL (1989) 7th Annual Membrane Technology/Planning Conference, Cambridge, MA, October 17–19, 1989
10. Sara M, Sleytr UB (1988) Membrane biotechnology: two-dimensional protein crystals for ultrafiltration. In: Rehn H-J, Reed G (eds.) Biotechnology: A comprehensive treatise in eight volumes, vol 6b, VCH, Germany
11. Furneau RC, Rigby UR, Davidson AP (1987) Porous fibers and method of forming them. US Patent 4,687,551. August 18, 1987
12. Thomas MP, Davidson AP, Butler C, Cole KS (1990) Synthesis and performance of new inorganic composite membranes. Proceedings ICOM, August 1990, Chicago, vol 1, p 528
13. Hennepe HJC, Bargeman D, Mulder MHV, Smolders CA (1987) J Memb Sci 35:39
14. Michaels AS (1990) Desal 77:5
15. Way JD, Nobel RD, Reed DL, Ginley GM, Jarr LA (1987) AIChE J 33:480
16. Belfort G (1989) J Memb Sci 40:123
17. Anderson MR, Mattes BR, Reiss H, Kaner RB (1991) Science 252:1412
18. Hagen DF, Craig S, Markell G, Schmitt G, Blevins D (1990) Anal Chima Acta 236:157
19. Stairmand JW, Bellhouse BJ (1985) Int J Heat Mass Transfer 27:1405
20. Kroner KH, Riesmeier B, Nissinen V, Kula MR (1986) Recent studies of dynamic filtration in enzyme recovery. Engineering Foundation Conference on Recovery of Bioproducts, Uppsala, May 1986, Sweden
21. Beyeler W, Thales T, Clement R (1989) Proceedings of 2nd Annual Meeting of Japanese Association for Animal Cell Technology, November 20–22, 1989, Tsukuba, Ibaraki, Japan
22. Davidson AP, Thomas MP, Azubike DC, Gallagher PM (1990) Proceedings of 5th World Filtration Conference, Nice, France. p 235
23. Edwards MF, Wilkinson WL (1971) Trans Inst Chem Engn 49:85
24. Sobey IJ (1980) J Fluid Mech 96:1
25. Winzeler H (1990) Poster presentation at the 5th European Biotechnology Conference, Copenhagen, Denmark
26. Kroner KH, Nissinen V, Ziegler H (1987) Bio/Technol 5:921
27. Ilias S, Govind K (1990) Sep Sci Tech 25:1307
28. Culkin B (1991) Abstract Book of the 4th National Meeting of the North American Membrane Society, San Diego, CA, May 29–31, paper 3B
29. Belfort GB, Weigand RJ, Mahar JT (1985) Particulate membrane fouling and recent developments in fluid mechanics of dilute suspensions. In: Sourirajan S, Matsuura T (eds) Reverse osmosis and ultrafiltration, ACS Symposium Series (vol 281) ACS, Washington
30. Garnick RL, Solli NJ, Papa PA (1988) Anal Chem 60:2546
31. Mattiasson G, Ramstorp W (1983) Ultrafiltration affinity purification. In: Biochemical engineering III, Annals of the New York Academy of Sciences, vol 413
32. Gudernatsch W, Krumbholz CH, Strathman H (1990) Desal 79:249
33. Heath CA, Belfort G (1990) Int J Biochem 22:823
34. Karel SF, Libicki SB, Robertson CR (1985) Chem Engng Sci 40:1321
35. Chang HN (1987) Biotechnol Adv 5:129
36. Belfort G (1989) Biotechnol Bioeng 33:1047

37. Cheryan M, Mehaia MA (1986) Membrane bioreactors In: McGregor WC (ed.) Membrane separations in biotechnology, Marcel Dekker, New York
38. King GA, Daugulis AJ, Faulkner P, Goosen MFA (1987) Biotechnol Prog 3:231
39. Sinacore MS, Creswick BC, Buehler R (1989) Bio/Technol 7:1275
40. Gharapetian H, Davies NA, Sun AM (1986) Biotechnol Bioeng 28:1595
41. Goosen MFA, O'Shea GM, Gharapetian HM, Chou S, Sun AM (1985) Biotechnol Bioeng 27:146
42. Lamberti FV, Evangelista RA, Blyszniuk J, Sefton MA (1984) Appl Biochem Biotech 10:101
43. Heath CA, Belfort G (1988) Adv Biotech Eng/Biotechnol 34:1
44. Custer L (1988) Physiological studies of hybridoma cultivation in hollow fiber bioreactors. PhD Thesis, University of California at Berkeley, CA
45. Rainen L (1988) Am Biotechnol Lab 6:20
46. Tharakan JP, Chau PC (1986) Biotechnol Bioeng 28:329
47. Feder J, Tolbert WR (1983) Scientific Amer 248:36
48. Gallagher SL, Tharakan JT, Chau PC (1987) Biotechnol Tech 1:91
49. Ozturk SS, Palsson BO, Midgley AR, Halberstadt CR (1989) Biotechnol Tech 3:55
50. Margaritis A, Wilke CR (1978) Biotechnol Bioeng 20:727
51. Märkl H, Kurosawa H, Niebuhr-Redder C, Kasche V (1990) New membrane bioreactor for the cultivation of animal cells. Proceedings ICOM, August 1990, Chicago, vol 2, p 968
52. Kornfield J, Stephanopoulos G, Voecks GE (1986) Biotechnol Prog 2:98
53. Schonberg JA, Belfort G (1987) Biotechnol Prog 3:80
54. Heath CA, Belfort G, Hammer BE, Mirer SD, Pimbley JM (1990) AIChE J 36:547
55. Hammer BE, Heath CA, Mirer SA, Belfort G (1990) Bio/Technol 8:327
56. Webster IA, Shuler ML (1978) Biotechnol Bioeng 20:1541
57. Webster IA, Shuler ML, Rony PR (1979) Biotechnol Bioeng 21:1725
58. Webster IA, Shuler ML (1981) Biotechnol Bioeng 23:447
59. Davis ME, Watson LT (1985) Biotechnol Bioeng 27:182
60. Davis ME, Watson LT (1986) Chem Eng J 33:133
61. Kleinstreuer C, Agarwal SS (1986) Biotechnol Bioeng 28:1233
62. Chresand TJ, Gillies RJ, Dale BE (1988) Biotechnol Bioeng 32:983
63. Salmon PM, Libicki SB, Robertson CR (1988) Chem Engng Comm 66:221
64. McKinney K, Dilwith R, Belfort G (1991) Biotech Prog (in press)
65. Ghose TK, Kostick JA (1970) Biotechnol Bioeng 12:921
66. Rony PR (1972) J Am Chem Soc 94:8247
67. Davis JC (1974) Biotechnol Bioeng 16:1113
68. Lewis W, Middleman S (1974) AIChE J 20:1012
69. Waterland LR, Robertson CR, Michaels AS (1975) Chem Eng Commun 2:37
70. Korous R, Olson A (1977) Biotechnol Bioeng 19:1
71. Ohlson I, Tragardh G, Hahn-Hägerdal B (1978) Biotechnol Bioeng 26:647
72. Kan JK, Shuler ML (1978) Biotechnol Bioeng 20:217
73. Roozen JP, Pilnik W (1979) Enz Microb Technol 1:22
74. Henley RG, Yang RYK, Greenfield PF (1980) Enz Microb Technol 2:206
75. Gekas VC (1986) Enz Microb Technol 8:450
76. Pizzichini M, Fabiani C, Adami A, Cavazzoni V (1989) Biotechnol Bioeng 33:955
77. Matson SL, Quinn JA (1986) Membrane reactors in bioprocessing. In Biochemical Engineering IV, New York Academy of Sciences, New York, vol 49
78. Ishikawa H, Tanaka T, Takase S, Hikita H (1989) Biotechnol Bioeng 34:357
79. Ishikawa H, Takase S, Tanaka T, Hikita H (1989) Biotechnol Bioeng 34:369
80. Ikemi M, Ishimatsu Y (1990) J Biotechnol 14:211
81. Howaldt M, Chmiel H, Kulbe KD (1990) NAD(H)-rejection in a charged enzyme membrane reactor. Proceedings ICOM, August 1990, Chicago, vol 2, p 957
82. Kise S, Hayashida M (1990) J Biotechnol 14:221
83. Ikemi M, Koizumi N, Ishimatsu Y (1990) Biotechnol Bioeng 36:149
84. Ikemi M, Ishimatsu Y, Kise S (1990) Biotechnol Bioeng 36:155
85. Cousineau J, Chang TMS (1977) Biochem Biophys Res Commun 79:24
86. Chang TMS, Malouf C (1978) Trans Am Soc Artif Intern Organs 24:18
87. Ilan E, Chang TMS (1986) Appl Biochem Biotechnol 13:221
88. Wahl HP, Chang TMS (1987) J Mole Catal 39:147
89. Gu KF, Chang TMS (1990) Appl Biochem Biotechnol 26:115

90. Jones CKS, Yang RYK, White ET (1988) AIChE J 34:293
91. Nakajima M, Watanabe A, Jimbo N, Nishizawa K, Nakao S (1989) Biotechnol Bioeng 33:856
92. Jeong YS, Vieth WR, Matsuura T (1991) Biotechnol Prog 7:130
93. Lopez JL, Wald SA, Matson SL, Quinn JA (1990) Ann NY Acad Sci 613:155
94. Ladner GM, Whitesides WE (1984) J Am Chem Soc 106:7250
95. Whitesides WE (1988) U.S. Patent No. 4,732,853
96. Wu D-R, Belfort G, Cramer SD (1990) I & EC Res 29:1612
97. Pronk W, Kerkhof PJAM, van Helden C, van't Riet K (1988) Biotechnol Bioeng 32:512
98. van der Padt A, Edema MJ, Sewalt JJW, van't Riet K (1990) J Amer Oil Chem Soc 67:347
99. Michaels AS, Matson SL (1985) Desal 53:231
100. Nagata N, Herouvis KJ, Dziewulski DM, Belfort G (1989) Biotechnol Bioeng 34:447
101. Shiloach J, Kaufman JB, Kelly RM (1986) Biotechnol Prog 2:230
102. Maiorella B, Dorin G, Carion A, Harano D (1991) Biotechnol Bioeng 37:121
103. Sheehan J, DeVane W, Shiloach J, Ha Y, Surette E, Weinstein M (1990) Bench and pilot scale recovery of animal cells grown in suspension culture using a plate-and-frame crossflow filter system. Proceedings ICOM, August 1990, Chicago, vol 1, p 610
104. Hanisch W (1986) Cell harvesting. In: McGregor WC (ed) Membrane separations in biotechnology, Marcel Dekker, New York
105. van Reis R, Leonard LC, Hsu CC, Builder SE (1991) Biotechnol Bioeng 38:413
106. Rolghigo P (1991) private communication
107. Flaschel E, Wandrey C, Kula M-R (1983) Adv Biochem Engng/Biotechnol 26:74
108. Blatt WF, Dravid A, Michaels AS, Nelsen L (1970) Solute polarization and cake formation in membrane ultrafiltration: causes, consequences and control techniques. In: Flinn JE (ed) Membrane science and technology, Plenum, New York
109. deFillipe RP, Goldsmith RL (1970) Application and theory of membrane processes for biological and other macromolecular solutions. In: Flinn JE (ed) Membrane science and technology, Plenum, New York
110. Porter MC (1972) Ind Eng Chem Prod Res Dev 11:234
111. Porter MC (1979) Membrane filtration. In: Schweitzer PA (ed) Handbook of separation techniques for chemical engineers, McGraw Hill, New York
112. Vilker VL, Colton CK, Smith KA (1981) AIChE J 27:632
113. Vilker VL, Colton CK, Smith KA (1981) AIChE J 27:637
114. Sirkar KK, Prasad R (1986) Protein ultrafiltration – some neglected considerations. In: McGregor WC (ed) membrane separations in biotechnology, Marcel Dekker, New York
115. Velacangil O, Howell JA (1980) Protein ultrafiltration: theory of membrane fouling and its treatment with immobilized proteases. In: Cooper A (ed) Ultrafiltration membranes and applications, Plenum, New York, p 217
116. Cheryan M, Merin U (1980) A study of the fouling phenomenon during ultrafiltration of cottage cheese whey. In: Cooper A (ed) Ultrafiltration membranes and applications, Plenum, New York, p 619
117. Higuchi A, Mishuma S, Nakagawa T (1991) J Memb Sci 57:175
118. Chan WKY, Belfort M, Belfort G (1991) J Biotechnol 18:225
119. Dorin G, Thomson J, Hanisch W (1990) Biotechnol Prog 6:494
120. McGregor WC, Weaver JF, Tansey SP (1988) Biotechnol Bioeng 31:38
121. Mani KN (1991) J Memb Sci 58:117
122. Milby KH, Steinmeyer DE, Tripodi MK (1990) Ion exchange hollow fiber membranes for protein purification. Proceedings ICOM, August 1990, Chicago, vol 1, p 625
123. Molinari R, Torres JL, Michaels AS, Kilpatrick PK, Carbonell RG (1990) Biotechnol Bioeng 36:572
124. Ding H, Yang M-C, Schisla D, Cussler EL (1989) AIChE J 35:814
125. Ding H, Cussler EL (1990) Biotechnol Prog 6:472
126. Lowe CR (1984) J Biotechnol 1:38
127. Zale SE, O'Connor JL, Matson SL (1990) Hollow fiber affinity membrane separations in downstream processing. Proceedings ICOM, August 1990, Chicago, vol 1, p 160
128. Parsi EJ (1976) Desal 19:139
129. Iwata H, Saito K, Furusaki S, Sugo T, Okamoto J (1990) Development of immobilized metal affinity hollow fiber membrane to adsorb proteins. Proceedings ICOM, August 1990, Chicago, vol 1, p 163
130. Serafica A, Belfort G (1991) Manuscript in preparation

131. Krause S, Kroner KH, Deckwer WD (1991) Biotechnol Tech 5:199
132. Dall-Bauman L, Ivory CF (1990) Protein separation via affinity-mediated membrane transport. In: Hamel J-F, Hunter JB, Sikdar SK (eds) Downstream Processing and Bioseparation. ACS Symposium Series (vol 419) ACS, Washington DC
133. Hill CL, Bartholomew R, Beidler D, David GS (1983) Biotechniques 1:14
134. Herak DC, Merrill EW (1989) Biotechnol Prog 5:9
135. Herak DC, Merrill EW (1990) Biotechnol Prog 6:33
136. Mattiasson B, Ramstorp M (1984) J Chromatogr 283:322
137. Mattiasson B, Ling TB (1986) Ultrafiltration affinity purification: A process for large-scale biospecific separations. In: McGregor WC (ed) Membrane separations in biotechnology, Marcel Dekker, New York
138. Luong JHT, Nguyen AL, Male KB (1987) Bio/Technology 5:564
139. Pungor E, Afeyan NB, Gordon NF, Cooney CL (1987) Bio/Technology 5:604
140. Mattiasson B, Ramstorp M (1983) Ann NY Acad Sci 413:307
141. Choe TB, Masse P, Verdier A (1986) Biotechnol Lett 8:163
142. Ting TGI, Mattiasson B (1989) Biotechnol Bioeng 34:1321
143. Sakoda A, Nigam SC, Wang HY (1990) Enz Microb Technol 12:349
144. Klein E (1991) Affinity Membranes, John Wiley, NY
145. Prasad R, Sirkar KK (1988) AIChE J 34:177
146. Prasad R, Sirkar KK (1990) J Memb Sci 50:153
147. Wald SA, Kessler SB, Lopez JL (1990) Characterization of large-scale membrane devices for diffusive transport. Proceedings ICOM, August 1990, Chicago, vol 1, p 41
148. Dahuron L, Cussler EL (1988) AIChE J 34:130
149. Prasad R, Sirkar KK (1989) J Memb Sci 47:235
150. Basu R, Sirkar KK (1990) Citric acid extraction with microporous hollow fibers. Proceedings ICOM, August 1990, Chicago, vol 1, p 641
151. Sengupta A, Basu R, Sirkar KK (1988) AIChE J 34:1698
152. Sørenson BV, Callahan RW (1990) Penicillin separations with contained liquid membranes. Proceedings ICOM, August 1990, Chicago, vol 1, p 695
153. Dordick JS, Rethwisch DG, Gao Y (1990) Enzyme-facilitated transport of organic acids through a liquid membrane. Proceedings ICOM, August 1990, Chicago, vol 1, p 673
154. Jeon YJ, Lee YY (1989) Enz Microb Technol 11:575
155. Cho T, Shuler ML (1986) Biotechnol Prog 2:53
156. Efthymiou GS, Shuler ML (1987) Biotechnol Prog 3:259
157. Steinmeyer DE, Shuler ML (1990) Biotechnol Prog 6:286
158. Scheper T, Hawachs W, Schügerl K (1984) Chem Eng J 29:B31
159. Makryaleas K, Scheper T, Schügerl K, Kula MR (1985) Germ Chem Eng 6:345
160. Scheper T, Likidis Z, Makryaleas K, Nowottny C, Schügerl K (1987) Enz Microb Technol 9:625
161. Meyer ER, Scheper T, Hitzmann B, Schügerl K (1988) Biotechnol Tech 2:127
162. Scheper T (1990) Adv Drug Del Rev 4:209
163. Jain SM (1983) Ann NY Acad Sci 413:290
164. Bing DH, DiDonno AC, Regan M, Strang CJ (1986) Blood plasma processing by electrodialysis. In: McGregor WC (ed) Membrane separations in biotechnology, Marcel Dekker, New York
165. Gavach C, Sandeaux R, Sandeaux J (1990) Electrotransport of amino acids through ion exchange membranes. Separation of amino acids by electrodialysis. Proceedings ICOM, August 1990, Chicago, vol 2, p 870
166. Brors A, Kroner KH (1990) Electro-crossflow filtration of microbial suspensions. Proceedings ICOM, August 1990, Chicago, vol 1, p 616
167. Grimshaw PE, Grodzinsky AJ, Yarmush ML, Yarmush DM (1989) Chem Engng Sci 44:827
168. Devereux N, Hoare M (1986) Biotechnol Bioeng 28:422
169. Devereux N, Hoare M, Dunnill P (1986) Biotechnol Bioeng 28:88
170. Devereux N, Hoare M, Dunnill P (1986) Chem Eng Commun 45:255
171. Bentham AC, Treton MJ, Hoare M, Dunnill P (1988) Biotechnol Bioeng 31:984
172. Davis RH (1991) In: Ho W, Sirkar KK (eds) chap 33, Membrane Handbook (in press)
173. Forman SM, DeBernardez ER, Feldberg RS, Swartz RW (1990) J Memb Sci 48:263
174. Krimpenfort P, Rademakers A, Eyestone W, van der Schans A, van den Broek S, Kooiman P, Kootwijk E, Platenburg G, Pieper F, Strijker R, de Boer H (1991) Bio/Technol 9:844
175. Vadgama P (1990) J Memb Sci 50:141

# Aqueous Two-Phase Systems for Biomolecule Separation

A. D. Diamond and J. T. Hsu*
Bioprocessing Institute, Department of Chemical Engineering,
Lehigh University, Bethlehem, PA 18015, USA

Over the past thirty years, aqueous polymer two-phase technology has evolved, both experimentally and theoretically, into a separation science with many useful applications in biomolecule purification and bioconversion. This paper summarizes the developments in the applications of aqueous two-phase systems to biotechnology. The main topics to be considered are the phase diagram and its characteristics, fundamentals of biomolecule partition, large-scale and multi-stage aqueous two-phase biomolecule purification, and extractive bioconversions. The first topic involves a discussion of the thermodynamics of aqueous polymer two-phase formation and how it is influenced by such factors as polymer molecular weight and concentration, temperature, and salt type and concentration. Next, the theoretical and experimental aspects of biomolecule partition in aqueous two-phase systems will be discussed in light of the factors which influence biomolecule partition: polymer concentration and molecular weight; temperature; salt type and concentration; the addition of charged, hydrophobic and affinity derivatives. Having reviewed the fundamentals of phase diagram formation and biomolecule partition, the next two topics are applications of aqueous two-phase technology. The first set of applications involve the large-scale extraction of proteins using one to three equilibrium stages and multi-stage purifications using countercurrent distribution, liquid-

---

* To whom all correspondence concerning this paper should be addressed.

Advances in Biochemical Engineering
Biotechnology, Vol. 47
Managing Editor:
© Springer-Verlag Berlin Heidelberg 1992

liquid partition chromatography and continuous countercurrent chromatography. The second application, and very promising area for future aqueous two-phase technology, is the extractive bioconversion which permits the simultaneous production and purification of a biomolecule.

## List of Symbols and Abbreviations

$$A = m_3\left(\alpha_1\left(\frac{1}{m_1} - 1 + \chi_{03} - \chi_{13} + \chi_{01}\right)\right.$$
$$\left. + \alpha_2\phi\left(\frac{1}{m_2} - 1 + \chi_{03} - \chi_{23} + \chi_{02}\right)\right)$$

$$A_1 = m_1\left(\alpha_1\left(\frac{1}{m_1} - 1 + 2\chi_{01}\right) + \alpha_2\phi\left(\frac{1}{m_2} - 1 + \chi_{01} + \chi_{02} - \chi_{12}\right)\right)$$

$$A_2 = m_2\left(\alpha_2\phi\left(\frac{1}{m_2} - 1 + 2\chi_{02}\right) + \alpha_1\left(\frac{1}{m_1} - 1 + \chi_{02} + \chi_{01} - \chi_{12}\right)\right)$$

$$A^* = A + \frac{z_b F g}{RT}$$

$$b = m_3(\chi_{02}\alpha_2^2\phi^2 - \chi_{01}\alpha_1^2)$$

$$b^* = b + \frac{z_b F h}{RT}$$

$c_i$ = concentration of species i, %(w/w)
$d$ = density of a phase, $g\,ml^{-1}$
$F$ = Faraday constant
$g$ = regression constant in Eq. (19)
$h$ = regression constant in Eq. (19)
$K_a$ = binding constant for protein-ligand complex
$K_i$ = $w_i''/w_i'$, partition coefficient for species i
$K_L$ = partition coefficient of biomolecule in the presence of polymer bound ligand
$K_{L,o}$ = partition coefficient of biomolecule in the absence of polymer bound ligand
$K_{3,o}$ = partition coefficient of biomolecule when $\Delta\psi = 0$ or $z_b = 0$
$K_+$ = partition coefficient of a cation
$K_-$ = partition coefficient of an anion
$k$ = Boltzman's constant
$L$ = ligand concentration, %(w/w)
$m_i$ = molar volume ratio of species i to that of water
$M$ = molarity, $mol\,l^{-1}$
$n$ = number of independent binding sites on a protein
$P_s$ = $c_s^i/c_s^{Dex}$, distribution coefficient

R     = gas law constant, (K-atm)/(mol-K)
T     = absolute temperature, K
V     = volume fraction of a phase, %(v/v)
W     = weight fraction of a phase, %(w/w)
$w_i$    = weight fraction of species i, %(w/w)
$Y_3$    = biomolecule yield, %
z     = lattice coordinate number
$z_b$    = charge of a biomolecule
$z^+$    = charge of a cation
$z^-$    = charge of an anion

## Greek Letters

$\alpha_i$    = proportionality factor between volume and weight fraction for species i
$\beta_s$    = regression constant in Eq. (8)
$\gamma^*$    = constant in Eq. (16)
$\Delta G_m$ = Gibbs free energy of mixing, $J\,mol^{-1}$
$\Delta H_m$ = enthalpy change on mixing, $J\,mol^{-1}$
$\Delta S_m$ = entropy change on mixing, $J\,mol\text{-}K^{-1}$
$\Delta w_{ij}$ = energy change for the formation of a contact between species i and j, J
$\Delta\psi$    = electrostatic potential difference between the phases, Volts
$\delta^*$    = constant in Eq. (16)
$\varepsilon^*$    = constant in Eq. (17)

$$\phi = \frac{w_2'' - w_2'}{w_1'' - w_1'}$$

$\chi_{ij}$    = Flory-Huggins interaction parameter between species i and j
$\psi$     = electrostatic potential, Volts

## Superscripts

$''$     = top phase
$'$     = bottom phase
DEX = dextran-rich phase
i     = i-rich phase

## Subscripts

1     = PEG in the PEG/dextran/water and PEG/potassium phosphate/water systems; dextran in the ficoll/dextran/water system
2     = dextran in the PEG/dextran/water system; potassium phosphate in the PEG/potassium phosphate/water system; ficoll in the ficoll/dextran/water system
3     = biomolecule
i     = component i
s     = salt

# 1 Introduction

The impetus for research and development in the field of bioseparations has been sparked by the difficulty and complexity in the downstream processing of pharmaceutical and biological products. Indeed, 50% to 90% of the production cost for a typical biological product resides in the purification strategy. There is a need for efficient, effective and economical large-scale bioseparation techniques which will achieve high purity and high recovery, while maintaining the biological activity of the molecule. One such purification technique which meets these criteria involves the partitioning of biomolecules between two or more immiscible phases in an aqueous system.

The multiphase aqueous system has been a topic of interest for some time, and was first reported in the literature by Beijerinck [1,2]. He discovered that when gelatin, agar and water were mixed at certain concentrations, an aqueous two phase system would result, the top phase being rich in gelatin, while the bottom phase rich in agar. In 1947, Dobry and Boyer-Kawenoki [3] performed a systematic study on the miscibility of pairs of polymers in the presence of water or organic solvents. They found that phase separation was rather commonplace. In the 1940s and 1950s, Craig and Craig [4] pioneered the use of organic/aqueous two phase systems for protein purification using countercurrent distribution. However, it was not until 1955 that Per-Åka Albertsson discovered that poly(ethylene glycol) (PEG), potassium phosphate and water, and then PEG, dextran and water formed two-phases [5]. The PEG/dextran/water and PEG/salt/water systems have since then been the most frequently investigated and utilized aqueous two-phase systems for biomolecule purification.

Aqueous two-phase systems offer different physical and chemical environments which allow for the partitioning of solutes such as proteins, cells, cell particles and nucleic acids [6]. The differences in the phases are small [6], and therefore preclude the harsh treatment offered by traditional extraction systems. For example, the phases of PEG/dextran/water systems contain between 80% and 99% water by weight, possess extremely low interfacial tensions (on the order of $10^{-7}$ N cm$^{-1}$), and have been shown to provide a protective environment for biological materials [6]. This compares rather favorably with organic solvent/water systems, such as butanol/water and ethanol/aqueous salt solutions pioneered by Craig and Craig [4]. Such systems have high interfacial tensions (on the order of $10^{-4}$ N cm$^{-1}$) and organic phases containing only 40% to 50% water. These conditions lead to several problems, including precipitation and denaturation of proteins, and concentration of biological materials exclusively in the aqueous phase. These deleterious effects can be avoided through the use of aqueous polymer two-phase systems.

Aqueous polymer two-phase systems are extremely powerful for the separation and analysis of biological particles. Over the past three decades, many researchers have applied these systems to laboratory-scale separation of proteins, cells, cell organelles, viruses, cell membrane fragments and other biological materials [6, 17, 18]. Various researchers have utilized these systems as a means of determining surface properties of biomolecules such as charge and hydrophobicity [19–24]. The application of aqueous polymer two-phase systems to large-scale purification of proteins has also been demonstrated using centrifuges as extractors [25–28]. The main advantages of using such systems have been summarized by Albertsson [6] and are given below:

1. Scale-up can easily and reliably be predicted from small laboratory experiments.
2. Rapid mass transfer and equilibrium is reached by relatively little input of energy in the form of mechanical mixing.
3. Continuous processing is readily achievable.
4. The polymers stabilize the enzymes.
5. Separation can be made selective and rapid.
6. Separation can be carried out at room temperature due to the rapid separation.
7. It has proven to be more economical than other separation processes.

In particular, the case of scale-up is demonstrated by the fact that the partition coefficient is essentially independent of the volume of the system [6]. Scale-up by a factor of 25 000 has readily been accomplished [29]. In order to facilitate the use of these systems for biomolecule purification, and bring these systems from laboratory to production scale, a need still exists for a simple and effective means for characterization of the aqueous biphasic systems, selection of the proper system for protein purification, and correlation of biomolecule partition coefficients. The purpose of this paper is to report on the current developments in these areas, and to report on the emerging biotechnical applications of the systems.

## 2 Aqueous Two-Phase Systems and Their Characterization

When pairs of water soluble polymers or a water soluble polymer and low molecular weight solute are mixed with water above critical concentrations, an aqueous two-phase system will form. An extensive list of these systems and their components has been developed by Albertsson [6] and is summarized and updated in Table 1. This table divides the systems into two categories: 1) polymer-polymer-water systems, and 2) polymer-low molecular weight component-water systems. The most commonly used systems are PEG/dextran/water and PEG/potassium phosphate/water.

Each two-phase phase system can be characterized by its unique phase diagram, which contains the equilibrium phase compositions for the system. The most fundamental data for any type of liquid-liquid extraction process are the

**Table 1.** Aqueous two-phase systems and their components

A. Polymer-Polymer-Water Systems

| Polymer | Polymer | Reference |
|---|---|---|
| Polypropylene Glycol | Methoxypolyethylene Glycol | [7] |
| | Polyethylene Glycol | [8] |
| | Polyvinyl Alcohol | [7] |
| | Polyvinylpyrrolidone | [8] |
| | Hydroxypropyl Dextran | [8] |
| | Dextran | [8] |
| | Hydroxypropyl Starch | [9] |
| | Maltodextrin | [10] |
| Polyethylene Glycol | Polyvinyl Alcohol | [8] |
| | Polyvinylpyrrolidone | [8] |
| | Dextran | [8] |
| | Ficoll | [7] |
| | Pullulan | [11] |
| Polyvinyl Alcohol | Methylcellulose | [12] |
| | Hydroxypropyl Dextran | [7] |
| | Dextran | [8] |
| Polyvinylpyrrolidone | Methylcellulose | [13] |
| | Dextran | [8] |
| Methylcellulose | Hydroxypropyl Dextran | [7] |
| | Dextran | [8] |
| Ethylhydroxyethylcellulose | Dextran | [8] |
| | Hydroxypropyl Starch | [14] |
| Hydroxypropyl Dextran | Dextran | [7] |
| Ficoll | Dextran | [7] |

B. Polymer-Low Molecular Weight Solute-Water

| Polymer | Low Molecular Weight Solute | Reference |
|---|---|---|
| Polypropylene Glycol | Potassium Phosphate | [8] |
| | Glucose | [8] |
| | Glycerol | [8] |
| Polyethylene Glycol | Potassium Phosphate | [8] |
| | Amino Acid | [15] |
| | Mono/Disaccharide | [15] |
| | Peptide/Protein | [15] |
| | Sodium Chloride | [16] |
| Methoxypolyethylene Glycol | Potassium Phosphate | [8] |
| Polyvinyl Alcohol | Butylcellosolve | [8] |
| Polyvinylpyrrolidone | Butylcellosolve | [8] |
| Dextran | Propyl Alcohol | [8] |

composition of the equilibrium phases. In the case of aqueous polymer two-phase systems, a comprehensive set of phase diagram data is needed in order to facilitate their use for biomolecule purification and to aid in the development of thermodynamic models for their prediction. Furthermore, such data are needed for the eventual prediction of biomolecule partition coefficients.

A significant amount of phase equilibrium data has been obtained for the PEG/dextran/water [6, 30–32] and PEG/potassium phosphate/water [6, 33] systems, and Albertsson [6] has recorded data for several of the other systems presented in Table 1. An example of a PEG 3400/Dextran T-500/water and PEG 3400/potassium phosphate/water phase diagram is presented in Fig. 1. The PEG 3400/Dextran T-500/water phase diagram (Fig. 1a) has been labeled at several points and will be used in the discussion that follows. The binodial curve is represented by curve containing the points DPC and contains the equilibrium phase compositions. The point P represents the plait point (critical point), which is the point at which the compositions and volumes of the two-phases theoretically become equal. Point A in the PEG 3400/Dextran T-500/water phase diagram is representative of the region to the left of the binodial curve where only one homogeneous phase is present. To the right of the binodial is point B. When a solution in this region is prepared, a two-phase system will result. The upper phase will have a composition represented by C and the lower phase will

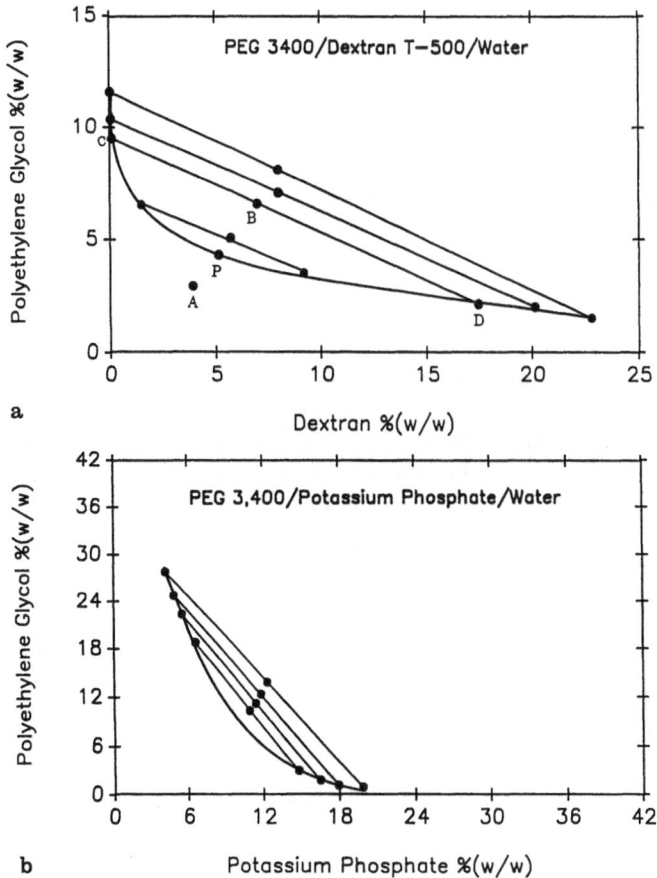

**Fig. 1a, b.** Aqueous polymer phase diagrams: **a)** PEG 3400/Dextran T-500/Water at 4 °C. Data from Diamond and Hsu [30]; **b)** PEG 3400/Potassium Phosphate/Water at 4 °C and pH 7.0. Data from Lei et al. [33]

have a composition represented by D. The ratio of the top phase weight percent to the bottom phase weight percent is represented by the ratio of the lengths of line segments $\overline{BD}$ and $\overline{BC}$, that is,

$$\frac{W''}{W'} = \frac{\overline{BD}}{\overline{BC}} \tag{1}$$

where W is the %(w/w) of the phase, and single and double prime superscripts refer to the bottom and top phases, respectively. The point B can be moved anywhere along the tie-line $\overline{CD}$ (which connects the equilibrium phase compositions) and the resulting phase system will always have top and bottom phase compositions given by points C and D, and the phase weight ratio will be defined by Eq. (1). If the density of the phases is known, then the volume ratio of the phases can be determined:

$$\frac{V''}{V'} = \frac{W''d'}{W'd''} = \frac{\overline{BD}\,d'}{\overline{BC}\,d''} \tag{2}$$

Since the densities of the two aqueous phases are generally very close to one another [6], the quantity $\overline{BD}/\overline{BC}$ can be used as an approximation for the phase volume ratio. As will be discussed later, the volume ratio is an important parameter when designing biomolecule purification strategies since a favorable (unfavorable) volume ratio can improve (reduce) biomolecule yield in a particular phase.

## 2.1 Phase Diagram Thermodynamics

Two important viewpoints exist on the driving force for phase separation in aqueous polymer two-phase systems. On the one hand several research groups [30, 34–36] have utilized the theory developed by Flory [37] and Huggins [38] to describe the thermodynamics which lead to phase separation. The simplest way to discuss their hypothesis is to first present the general expression for the Gibbs free energy of mixing at constant temperature and pressure:

$$\Delta G_m = \Delta H_m - T\Delta S_m \tag{3}$$

where $\Delta G_m$, $\Delta H_m$, $\Delta S_m$ and T refer to the Gibbs free energy of mixing, the enthalpy of mixing, the entropy of mixing, and absolute temperature, respectively. If $\Delta G_m$ is negative when the two polymers are mixed with water, a solution will result without phase separation. If $\Delta G_m$ is positive, phase separation will occur. The Flory-Huggins theory suggests that if polymer solution concentrations are low, then there will only be a small gain in entropy upon mixing the polymers in water. However, since polymer chains have a much higher surface area per molecule than do low molecular weight compounds, the interaction between segments of the two polymer molecules, which are generally unfavorable, will lead to a positive $\Delta H_m$ which will dominate the $\Delta G_m$ expression [39]. The positive $\Delta G_m$ will result in phase separation. Therefore, it can be concluded that by using the Flory-Huggins theory, it is assumed that water does not play a key role in determining the phase separation, rather it is the polymer interaction.

On the other hand, Zaslavsky et al. [40] have advocated that the structure of water should also be considered when discussing phase separation. Their reasoning was based on the fact that they found the physico-chemical properties of water (dielectric relaxation time, static dielectric constant, relative affinity of water for a $CH_2$ group and overall polarity) in the coexisting phases of the systems were different [40]. In addition, it was found that changing different factors (temperature, organic solvents, inorganic salts) produced similar effects on aqueous two-phase diagram behavior. The only thing the factors have in common is that they affect the structure of water.

In order to facilitate the use of aqueous two-phase systems and provide a basis for selecting a system for biomolecule purification, a sound thermodynamic theory is needed for prediction and correlation of phase equilibrium behavior. Several theoretical models have been proposed for the thermodynamic behavior of aqueous two-phase systems [41–48]. Diamond [47] and Diamond and Hsu [48] developed a correlation for aqueous polymer phase diagram behavior based on the Flory-Huggins theory. The Flory-Huggins theory, though simplistic in nature, represents the classical approach for describing the thermodynamics of phase separation in polymer systems. According to this theory, the polymers are linear (monodisperse), long chained random coils, while the solvent is monomeric. The geometry of the polymer and solvent are considered essentially identical. The polymer and solvent molecules exist in the form of a single lattice scheme, each cell of which may be occupied by either a solvent molecule or the segment on a linear polymer. The entropy change on mixing is a combinatorial term reflecting the variety of ways of arranging the polymers and solvent in the lattice, while the heat of mixing represents the energy change associated with the formation of contacts between unlike neighbors in the lattice.

In order to apply the Flory-Huggins theory of polymer solution thermodynamics to aqueous two-phase systems, several fundamental assumptions must be made. These include 1) the liquid lattice model provides an adequate representation of an aqueous solution, 2) an ideal entropy of mixing is suitable, 3) deviations from ideal solution behavior can be simply accounted for in terms of (enthalpic) solute-solvent and solute-solute interaction parameters, and 4) the polymer chains are fully flexible. Inherent in the first three assumptions is that molecular interactions are almost exclusively of the van der Waals type. However, aqueous two-phase systems will undoubtedly possess, among others, hydrogen bonding and some ionic interactions.

These relationships developed based on the Flory-Huggins theory are as follows [47, 48]:

$$\ln(K_1) = A_1(w_1'' - w_1') \tag{4}$$

and

$$\ln(K_2) = A_2(w_1'' - w_1')$$ (5)

where,

$$A_1 = m_1\left(\alpha_1\left(\frac{1}{m_1} - 1 + 2\chi_{01}\right) + \alpha_2\phi\left(\frac{1}{m_2} - 1 + \chi_{01} + \chi_{02} - \chi_{12}\right)\right)$$ (6)

$$A_2 = m_2\left(\alpha_2\phi\left(\frac{1}{m_2} - 1 + 2\chi_{02}\right) + \alpha_1\left(\frac{1}{m_1} - 1 + \chi_{02} + \chi_{01} - \chi_{12}\right)\right)$$ (7)

$K_i = w_i''/w_i'$, $w_i$ is the weight fraction of species i, $m_i$ is the molar volume of species i to that of water, $\chi_{ij}$ is the Flory-Huggins interaction parameter for species i and j, $\alpha_i$ and $\phi$ are constants, the single and double prime superscripts refer to the bottom and top phase, respectively, PEG is defined as species 1 and dextran is species 2, and the quantity $(w_1'' - w_1')$ is referred to as the PEG concentration difference between the phases. The use of Eqs. (4) and (5) is illustrated in Fig. 2 for the PEG 3400/Dextran T-500/water system at 4 °C. The phase diagram for this system was presented in Fig. 1a. The binodial is represented by the two straight lines which intersect at (0,0), the plait point, labeled P. The open and closed symbols represent the ln $(K_i)$ for PEG and dextran, respectively. At a particular value of the PEG concentration difference, there will be an ln $(K_1)$ and ln $(K_2)$ which correspond to the phase equilibrium composition for a tie-line on the phase diagram. For example, the points C and D which connect a tie-line in Fig. 1a have been indicated in Fig. 2. The simplicity

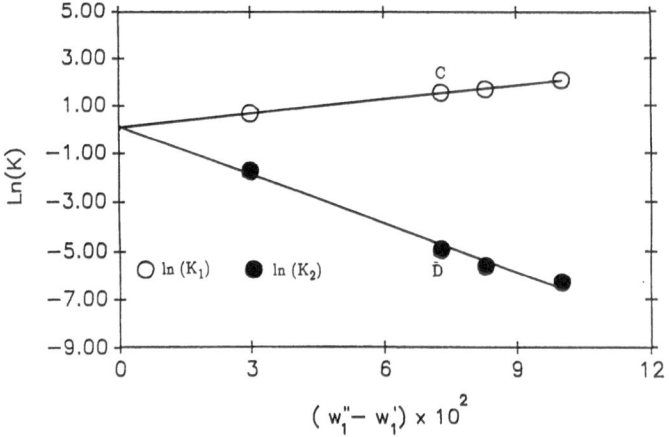

**Fig. 2.** Correlation of aqueous polymer phase diagram data using Eqs. (4) and (5). The system is PEG 3400/Dextran T-500/water at 4 °C. Points C and D correspond to Fig. 1a. Adapted from Diamond [47]

of using Eqs. (4) and (5) is based on the fact that only a single phase equilibrium composition ($w_1''$, $w_2''$, $w_1'$, $w_2'$) needs to be measured in order to determine $A_1$ and $A_2$ and thus, determine the phase compositions of the phase system. Therefore, when working with a new phase system, extensive phase equilibrium determination can be avoided. The equation has been proven to be applicable to PEG/dextran, Ficoll/dextran and PEG/salt aqueous systems [47]. With the ability to correlate phase system data, the next step is to consider the factors which effect the phase diagram.

## 2.2 Factors Influencing Phase Diagram Behavior

Aqueous polymer phase diagram can be influenced by many factors including polymer concentration and molecular weight, temperature, salt and pH. Although the effects and mechanisms by which they influence phase separation are still not completely understood, each will be discussed below.

### 2.2.1 Polymer Concentration

The effect of polymer concentration was briefly discussed with the description of the phase diagrams in Fig. 1. Generally, at low polymer concentrations, as in the region of point A in Fig. 1, each polymer is miscible with water and with each other. As the polymer concentration is increased into the two-phase region (such as point B) phase separation will occur [6].

### 2.2.2 Polymer Molecular Weight

Polymer molecular weight will influence the polymer concentrations required for phase separation and the phase diagram symmetry. Generally, it has been observed for the PEG/dextran system that increasing polymer molecular weight will lead to lower polymer concentrations required for phase separation [6, 30–32]. In addition, as the difference in molecular size between the two polymers is increased, the phase diagram becomes more asymmetric [6].

### 2.2.3 Temperature

Decreasing the temperature of a PEG/dextran system will lead to lower polymer concentrations required for phase separation [6]. Similar results have been reported for the PVP/dextran and PVA/dextran systems [40]. However, the PEG/salt systems behave in an opposite manner. Decreasing temperature leads to higher polymer concentrations for phase separation [6].

*2.2.4 Salt Type and Concentration*

In PEG/dextran systems it has been shown that most salts composed of univalent ions (e.g. NaCl) have partition coefficients close to one [49, 50]. Zaslavsky et al. [51–53] have demonstrated that increasing concentrations of the univalent salts (up to 0.1 M) in PEG/dextran systems will alter the composition of the phases without significant effect upon the position of the binodial. However, multivalent salts, such as phosphate, sulfate and tartrate show an increasing tendency to partition to the bottom, dextran-rich phase with increasing salt concentration and distance from the critical point [49, 53]. These salt will significantly alter both phase composition and binodial position of the PEG/dextran system. Generally, the binodial is shifted to lower polymer concentration. The fact that multivalent salts are rejected to the bottom phase is undestandable since these salts will form two phases themselves with PEG at higher concentrations of both species [7].

The effect of salt type and concentration on PEG/dextran systems can not be generalized to other polymer-polymer-water systems. For example, Zaslavsky et al. [51–53] found that the salts KSCN, KI, KBr, KCl, $KNO_3$, KF, and $K_2SO_4$ shifted the PEG/dextran binodial to lower phase compositions as salt concentration was increased over the range 0.0 M to 1.0 M. The Ficoll/dextran binodial showed an initial shift to higher phase forming polymer concentrations and then shifted to lower phase forming concentrations as salt concentration was increased over the same range. The only exceptions were KF and $K_2SO_4$ which shifted the binodial to lower concentrations. The PVP/dextran system proved to be the very sensitive to salt type and concentration [52]. The salts KSCN, KI, KBr, $KNO_3$ and KCl shifted the binodial to higher polymer concentrations, while KF and $K_2SO_4$ initially shifted the binodials to higher concentrations and then sharply dropped off to lower polymer concentrations as salt concentration was increased.

In analyzing the effect of salt type and concentration on the PEG/dextran, PVP/dextran and Ficoll/dextran phase diagrams, Zaslavsky et al. [53] found that the salt composition of the phases was influenced by the polymer concentration of the system. The following empirical expression which relates the salt partition coefficient to the polymer concentration difference between the phases was developed [53]:

$$\ln P_s = \beta_s(c_i^i - c_i^{Dex}) = \beta_s \Delta c_i \qquad (8)$$

where $P_s = c_s^i/c_s^{Dex}$, $c_s$ and $c_i$ are the concentrations of salt and polymer i in a given phase, respectively, and the superscripts i and Dex denote the i-rich and dextran-rich phases, respectively. This notation was adopted since dextran is enriched in the bottom phase in the PEG/dextran system and in the top phase in the Ficoll/dextran system. It should be noted that Eq. (8) for salt partitioning is very similar to Eqs. (4) and (5) which were later developed by Diamond [47] and Diamond and Hsu [48] for phase forming polymer partitioning based on Flory-Huggins theory. Equation (8) was found to be valid for eight salts (NaSCN,

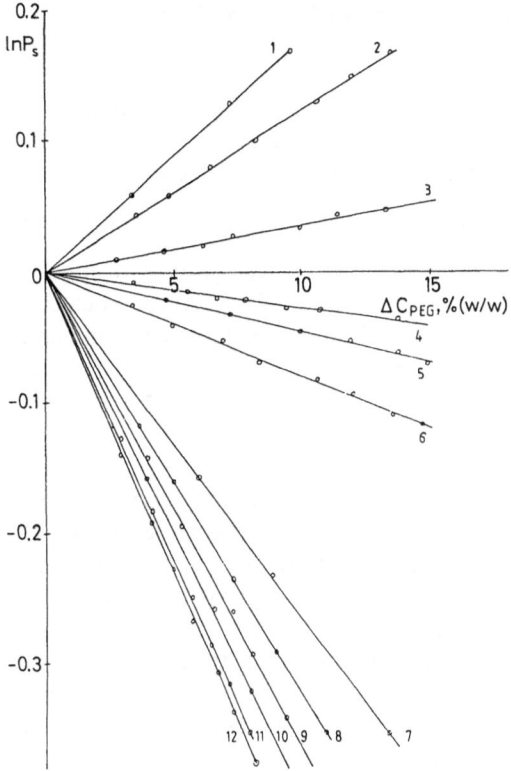

**Fig. 3.** Distribution coefficient ($P_s$) of salts in a PEG/Dextran/Water system as a function of the PEG concentration difference between the phases: $1 = NaSCN$ $(0.10\,mol\,kg^{-1})$; $2 = KSCN$ $(0.10–0.75\,mol\,kg^{-1})$; $3 = NH_4SCN$ $(0.10\,mol\,kg^{-1})$; $4 = KCl$ $(0.10\,mol\,kg^{-1})$; $5 = KCl$ $(0.50\,mol\,kg^{-1})$; $6 = KCl$ $(0.75\,mol\,kg^{-1})$; $7 = Na_2SO_4$ $(0.10\,mol\,kg^{-1})$; $8 = K_2SO_4$ $(0.05\,mol\,kg^{-1})$; $9 = (NH_4)_2SO_4$ $(0.10\,mol\,kg^{-1})$; $10 = Cs_2SO_4$ $(0.10\,mol\,kg^{-1})$; $11 = K_2SO_4$ $(0.10$ $mol\,kg^{-1})$; $12 = K_2SO_4$ $(0.25\,mol\,kg^{-1})$. From Zaslavsky et al. [53]

KSCN, $NH_4SCN$, KCl, $Na_2SO_4$, $(NH_4)_2SO_4$, $Cs_2SO_4$ and $K_2SO_4$) all at a concentration of 0.10 M. The $r^2$ value for the relationship was 0.983, and an example of a typical plot for the eight salts partitioned in the PEG/dextran system is given in Fig. 3. Zaslavsky et al. [53] also found in the three systems studied that the water-structure breaking salts ($NH_4SCN$, NaSCN, KSCN and $KClO_4$) favor the PEG, ficoll or PVP-rich phases while the water structure-making salts (KF, $(NH_4)_2SO_4$, $Cs_2SO_4$, $Na_2SO_4$ and $K_2SO_4$) favor the dextran-rich phase.

The effect of pH on phase separation has been explored by Lei et al. [33] for the PEG/potassium phosphate system. It was found that as pH was increased from 6 to 9.2, the binodial curve was shifted to lower polymer and salt concentrations.

## 3  Biomolecule Partitioning in Aqueous Two-Phase Systems

The partition coefficient, K, of a biomolecule in an aqueous two-phase system has been found to be a function of many variables. Albertsson [6] proposed the following relationship for complex macromolecules such as proteins:

$$\ln K = \ln K^\circ + \ln K_{elec} + \ln K_{hfob} + \ln K_{size} + \ln K_{biosp} + \ln K_{conf} \quad (9)$$

where elec, hyfob, size, biosp, conf refer to the electrochemical, hydrophobic, size, biospecific and conformational contributions to the partition coefficient from both the protein structural properties and the surrounding environmental conditions of the system, and $K^\circ$ includes other factors. Therefore, the protein partition coefficient may also be expressed as:

$$\ln K = \ln K_{environment} + \ln K_{structure} \quad (10)$$

The environmental conditions influencing protein partitioning include: 1) salt type and concentration, 2) pH, 3) phase forming polymer type, molecular weight and concentration, 4) the presence of polymer derivatives such as charged, hydrophobic or affinity types, 5) temperature, and 6) gravity. The structural properties of the protein include its molecular weight (size), primary, secondary, tertiary and quaternary structure, net charge, hydrophobicity and other surface properties. In order to predict biomolecule partitioning in aqueous two-phase systems, a fundamental understanding (both experimental and theoretical) of each of the above factors is ultimately needed.

### 3.1  Effect of Phase Forming Polymer Concentration

The effect of phase forming polymer concentration on protein partition can be more readily visualized by keeping in mind the phase diagrams presented in Fig. 1. At the plait point, the partition coefficient of biomolecules is 1.0. As polymer concentration is increased, thus, moving further away from the plait point, protein partitioning generally tends to become more extreme, i.e., the partition coefficient becomes exceedingly greater or less than 1.0 [6]. However, exceptions to this trend do exist. A protein's partition coefficient may first increase, reach a maximum, and then decrease as polymer concentration is decreased [6].

Several theoretical models have been proposed for the thermodynamic behavior of protein partitioning in aqueous two-phase systems and have generally involved relating the natural logarithm of the biomolecule partition coefficient to the phase forming polymer concentrations and the interactions among the species present in solution. Brooks et al. [34] and Albertsson [35] have shown that the Flory-Huggins theory could be used to qualitatively predict protein partition trends. Gustafsson and Wennerström [41] attempted to predict a binodial of the PEG/dextran aqueous two-phase system using Flory-

Huggins theory, while Kang and Sandler [44] attempted to analyze PEG/ dextran binodials using both the Flory-Huggins theory and the UNIQUAC equation [54]. Baskir et al. [42] have modified the theory of Scheutjens and Fleer [55, 56] to predict protein partitioning, while King et al. [43] extended the solution theory of Edmond and Ogston [57] by taking into account the electrostatic potential difference between the phases. Cabezas [45] and Forciniti and Hall [46] have applied Hill's [58] constant pressure solution theory to aqueous two-phase partition. The advantages and disadvantages of the above models have been discussed in detail by Baskir et al. [59] and Abbot et al. [60]. Abbot et al. [60] have also presented a molecular thermodynamic description betwen globular proteins and flexible non-ionic polymers which accounts for the changing nature of the polymer phase from a polymer coil to a net of polymers, as polymer molecular weight is increased. However, despite the extensive research efforts described above, a general methodology for correlation and prediction of protein partition coefficients, which is needed in order to facilitate the use of aqueous two-phase systems, has yet to be realized.

Diamond and Hsu [36, 61] developed the following relationship which was based on Flory-Huggins solution thermodynamics:

$$\frac{\ln(K_3)}{(w_1'' - w_1')} = A^* + b^*(w_1'' - w_1') \tag{11}$$

where,

$$A^* = A + \frac{z_b F g}{RT} \tag{12}$$

$$b^* = b + \frac{z_b F h}{RT} \tag{13}$$

$$A = m_3 \left( \alpha_1 \left( \frac{1}{m_1} - 1 + \chi_{03} - \chi_{13} + \chi_{01} \right) \right.$$
$$\left. + \alpha_2 \phi \left( \frac{1}{m_2} - 1 + \chi_{03} - \chi_{23} + \chi_{02} \right) \right) \tag{14}$$

$$b = m_3 (\chi_{02} \alpha_2^2 \phi^2 - \chi_{01} \alpha_1^2) \tag{15}$$

Equation (11) is an expression for correlating protein partitioning in aqueous two-phase systems, where the intercept $A^*$ is a function of protein and phase forming polymer molecular weight, and the protein-water, protein-polymer, polymer-water interaction parameters, pH and salt type and concentration. Similarly, the slope $b^*$ is a function of protein molecular weight, the polymer-water interaction parameters, pH and the salt type and concentration. The parameter $z_b$ refers to the charge of the biomolecule, F is Faraday's constant, R is the ideal gas law constant, T is absolute temperature, and g and h are constants which are a function of the electrostatic potential difference between the phases. The electrostatic potential will be discussed in detail in the section covering the effect of salt on protein partitioning. In order to verify the applicability of Eq. (11) for correlation of partition data, proteins were partitioned in four systems of the PEG 8000/Dextran T-500/water phase diagram at 4 °C, the PEG 3400/potassium phosphate/water system at 20 °C, and the Ficoll 400/Dextran T-500/water system at 23 °C [36, 47]. Results for each of the systems are presented in Figs. 4 to 6, respectively.

In order to provide further proof of the applicability of the correlation, it was applied to other data obtained from the literature. In Fig. 7 the partition data

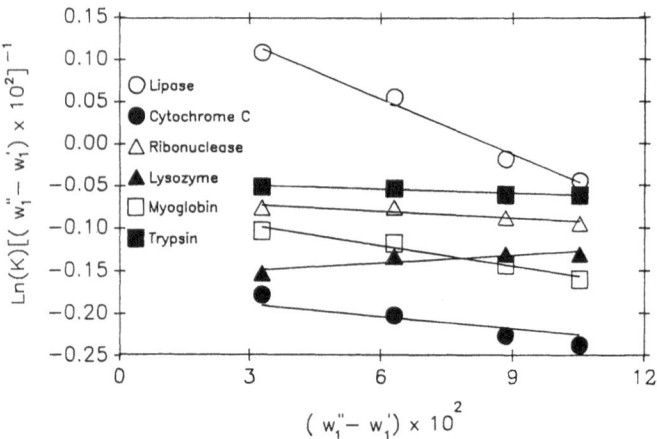

Fig. 4. Correlation of protein partition data according to Eq. (11) in the PEG 8000/Dextran T-500/Water systems at 4 °C, 0.01 molal Potassium Phosphate Buffer, pH 7.0. From Diamond and Hsu [36]

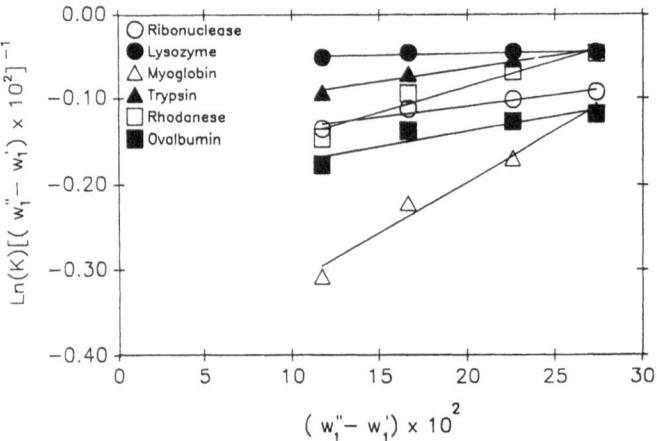

Fig. 5. Correlation of protein partition data according to Eq. (11) in the PEG 3400/Potassium Phosphate/Water systems at 20 °C, pH 7.0. From Diamond [47]

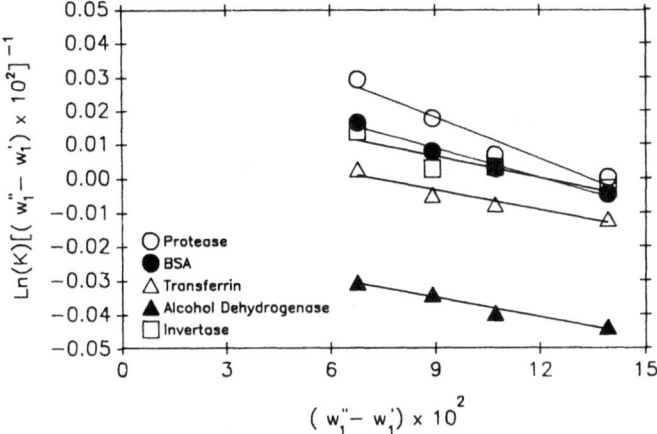

**Fig. 6.** Correlation of protein partition data according to Eq. (11) in the Ficoll 400/Dextran T-500/Water systems at 23 °C, 0.01 modal Potassium Phosphate Buffer, pH 7.0. From Diamond [47]

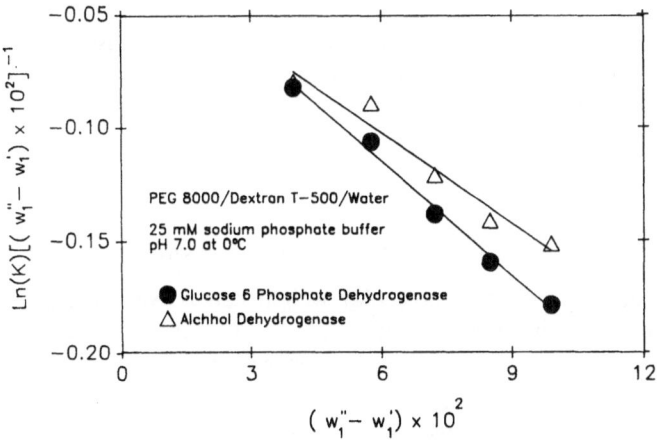

**Fig. 7.** Correlation of protein partition data from Johansson and Andersson [62] according to Eq. (11). From Diamond and Hsu [36]

of Johansson and Andersson [62] for the proteins glucose-6-phosphate dehydrogenase and alcohol dehydrogenase are presented. The system conditions are 25 mM sodium phosphate buffer, pH 7.0 and 0 °C. It is apparent that these two proteins follow the linear trend when correlated according to Eq. (11). It is also interesting to note that the two-phase system used in their study contained more than double the salt concentration compared with the partition data in Figs. 4 to 6. This indicates that the correlation can be applied to systems containing appreciable quantities of salt. In a similar fashion, the protein partition data from King et al. [43] has also been plotted according to Eq. (11) and are

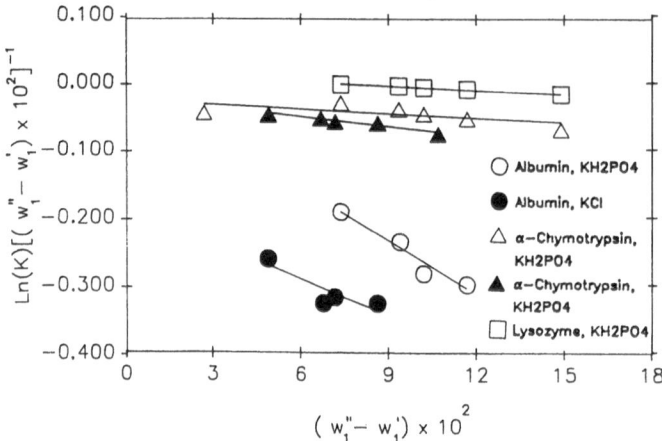

**Fig. 8.** Correlation of protein partition data from King et al. [43] according to Eq. (11). All partition data are at 25 °C. *Open* and *closed circles* represent partitioning in the PEG 3350/Dextran T-70/Water and PEG 8000/Dextran T-500/Water systems, respectively. From Diamond and Hsu [36]

presented in Fig. 8. The data from King et al., like that of Johansson and Andersson, show a linear trend. It should also be noted that King's data contained salt concentrations five times as great as the partition data in Figs. 4 to 6, indicating, once again, the applicability of Eq. (11) in salt-containing two-phase systems.

## 3.2 Effect of Phase-Forming Polymer Molecular Weight

The effect of phase forming polymer molecular weight can be very useful in manipulating a protein's partition coefficient [6, 25, 35]. Generally, increasing the molecular weight of a phase forming polymer will cause a protein to partition more towards the phase opposite to that in which the polymer is enriched. Similarly, if a phase forming polymer's molecular weight is decreased, a protein will tend to partition towards the phase in which that phase forming polymer is enriched. Therefore, in a PEG/dextran/water aqueous two-phase system, when dextran molecular weight is increased, a protein's partition coefficient will tend to increase since dextran is enriched in the bottom phase. Similarly, if dextran molecular weight is decreased, the protein will partition more into the bottom phase. These rules are also true for mixtures of proteins [63].

Hustedt et al. [25] showed that the partition coefficient of pullulanase decreased from approximately 1.3 to 0.1 as PEG molecular weight was increased from 1,000 to 40,000 in a system composed of 12% (w/w) PEG/1% (w/w) Dextran T-500/87%(w/w) water, 10 mM sodium phosphate buffer, pH 7.5. Albertsson et al. [35] demonstrated that the effect exerted by altering the phase forming polymer molecular weight also depends on the molecular weight of the protein. They showed that the magnitude of change in a protein's partition coefficient was small for proteins of molecular weight 10,000, but increased linearly up to molecular weights of 250,000.

When analyzing protein partition coefficients in aqueous two-phase systems which vary in phase forming polymer molecular weight, it must also be recognized that phase composition can also change as polymer molecular weight is changed. Albertsson et al. [35] apparently avoided this problem by using a system composed of 6% PEG/8% Dextran T-500. The composition of the equilibrium phases for this system have essentially the same composition regardless of phase forming polymer molecular weight [6].

An alternative method to avoid problems with phase composition when comparing protein partition coefficients in systems with different polymer molecular weights is to utilize the thermodynamic relationship expressed by Eq. (11). Diamond [47] partitioned bovine serum albumin (BSA) in four PEG/dextran systems which varied the dextran molecular weight from 40,000 to 500,000 and the PEG molecular weight from 3400 to 8000. The systems were as follows: PEG 8000/Dextran T-40, PEG 8000/Dextran T-70, PEG 8000/Dextran T-500 and PEG 3400/Dextran T-500. PEG 8000 and PEG 3400 have average molecular weights of 8,000 and 3,400, respectively, while Dextran T-40, Dextran T-70 and Dextran T-500 have weight average molecular weights of 40 000, 70 000 and 500 000, respectively. The BSA partition data are recorded in Fig. 9 according to the relationship of Eq. (11). At constant $(w_1'' - w_1')$, as dextran molecular weight is increased from 40 000 to 500 000 and PEG molecular weight is maintained constant at 8000, the quantity $\ln(K_3)/(w_1'' - w_1')$ increases and thus, the partition coefficient increases. This is in agreement with the observation cited above since dextran is enriched in the bottom phase and an increase in dextran molecular weight results in a higher partition coefficient. It can also be seen that as PEG molecular weight is increased from 3400 to 8000 (dextran

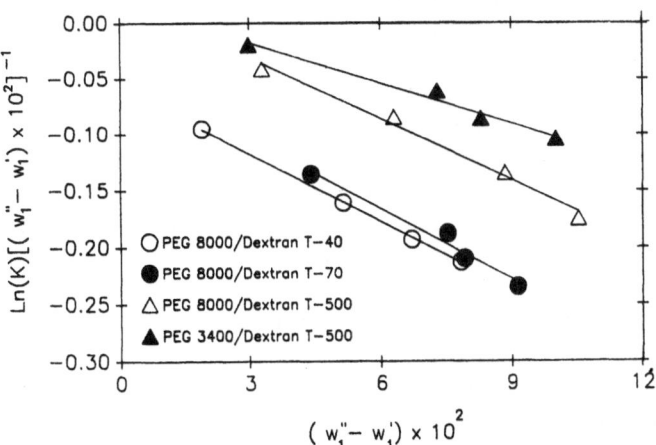

**Fig. 9.** Effect of PEG and Dextran molecular weight on BSA partitioning in PEG/Dextran/Water systems at 4 °C, 0.01 molal Potassium Phosphate Buffer, pH 7.0. From Diamond [47]

molecular weight constant at 500 000) the partition coefficient decreases at constant $(w_1'' - w_1')$.


## 3.3 Effect of Temperature

The effect of temperature on protein partitioning has not yet been thoroughly investigated, but some trends are apparent. Johansson et al. [64] partitioned phosphofructokinase from baker's yeast in a system composed of 7% Dextran T-500, 5% PEG 8000, 25 mM sodium acetate, 10 mM sodium phosphate buffer, pH 7.0, 5 mM 2-mercaptoethanol, 0.5 mM EDTA. As the temperature was increased from 0 °C to 40 °C the partition coefficient steadily increased. Johansson and Andersson [62] observed similar results for glucose-6-phosphate dehydrogenase and 3-phosphoglycerate kinase in the system composed of 7% Dextran T-500, 5% PEG 8000, 25 mM sodium phosphate buffer, pH 7.0. The partition coefficients for these two proteins tended to increase and then slightly level off as the temperature was increased over the 0 °C to 40 °C range.

It should be pointed out that an aspect not considered in the above partition experiments is the change in phase composition that will occur in a given phase system when temperature is changed. Generally, when temperature is increased for a PEG/dextran/water system, the binodial is shifted to higher phase forming concentrations [6], and the length of the tie-lines slightly decrease [6, 32]. Johansson et al. [65] apparently took this into consideration when they studied the partition of synaptic membranes, acetylcholinesterase, and succinate dehydrogenase in PEG/dextran systems over the temperature range of − 10 °C to 20 °C. The low temperature was obtained by adding organic solutes such as ethylene glycol or glycerol to the system. The change in phase composition described above was apparently handled by selecting a system for each temperature which had the same 'distance' from their respective binodial. The membranes and proteins partitioned to the interface, top phase and bottom phase. Generally, as temperature was increased over the − 10 °C to 20 °C range, the membranes and proteins increasingly partitioned to the interface and decreasingly partitioned to the bottom phase.

Equation (11) can be used to correlate the effect of temperature on protein partition coefficients using $(w_1'' - w_1')$ as the parameter for comparing different phase systems. Diamond and Hsu [36] and Diamond [47] partitioned a variety of proteins in the PEG/Dextran T-500 systems at 0 °C, 4 °C, 10 °C, and 22 °C. The results for lysozyme are presented in Fig. 10. At constant $(w_1'' - w_1')$, as temperature is increased, the lysozyme partition coefficient also is increased. This is in agreement with trends discussed earlier.

Diamond and Hsu [36] also derived relationships for the effect of temperature on the A* and b* parameters of Eq. (11):

$$A^* = \gamma^* + \frac{\delta^*}{T} \tag{16}$$

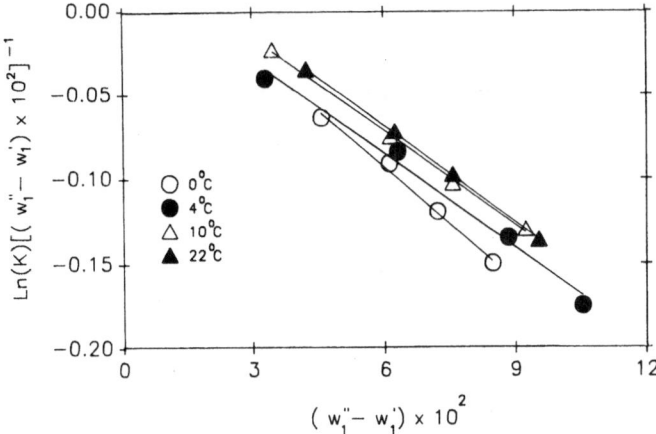

**Fig. 10.** Partition coefficient of Lysozyme in the PEG 8000/Dextran T-500/Water system at 0 °C, 4 °C, 10 °C, and 22 °C, 0.01 molal Potassium Phosphate Buffer, pH 7.0. From Diamond [47]

and

$$b^* = \frac{\varepsilon^*}{T} \tag{17}$$

where $\gamma^*$, $\delta^*$ and $\varepsilon^*$, are constants. In order to verify the effect of temperature on A* and b* the proteins cytochrome c, ribonuclease, lysozyme, myoglobin, trypsin, and BSA were partitioned the PEG 8000/Dextran T-500 systems at 0 °C, 4 °C, 10 °C, and 22 °C. In Fig. 11, A* is plotted versus (1/T) according to Eq. (16); [36, 47]. The data indicate a linear trend suggesting that Eq. (16) is useful for summarizing the variation of A* over the given temperature range. It is interesting to note that, for each of the proteins in Fig. 11, the data point at

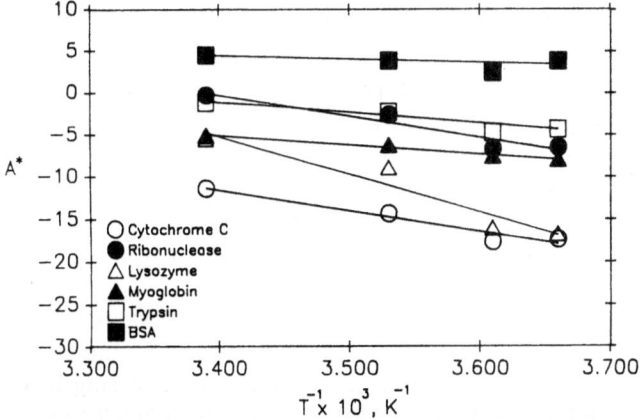

**Fig. 11.** The parameter A* as a function of temperature based on Eq. (16). From Diamond and Hsu [36]

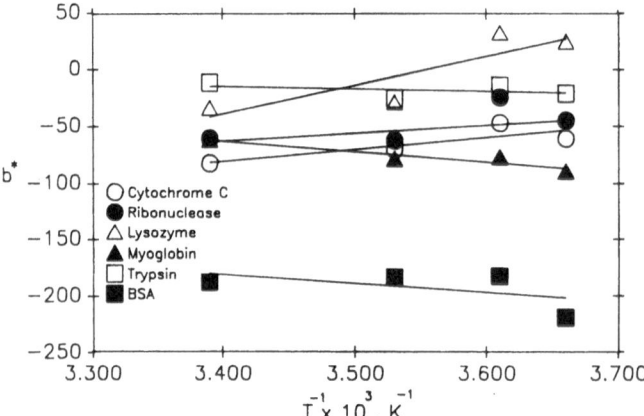

**Fig. 12.** The parameter b* as a function of temperature based on Eq. (17). From Diamond and Hsu [36]

4 °C consistently provides the greatest deviation from the linear trend. Although there is no definite explanation for this, it may be due to density changes occurring at the different temperatures. In Fig. 12, b* is plotted versus (1/T) according to Eq. (17). Although the data does not perfectly obey Eq. (17), i.e. at high temperatures b* should approach zero, the linear trend is still apparent. This deviation may be due to the assumptions and simplifications inherent in the thermodynamic model.

## 3.4 Effect of Salt

The addition of salts to aqueous two-phase systems can be used rather effectively to alter the partition coefficient of proteins. At low salt concentrations (0.1 to 0.2 M) and proteins far away from their isoelectric points the effects of salt type and concentration can be dramatic [66–73]. Generally speaking, for negatively (positively) charged proteins in PEG/dextran aqueous systems, the partition coefficient will be successively decreased (increased) in the series sulfate > floride > acetate > chloride > bromide > iodide    and    lithium > ammonium > sodium > potassium [63]. For proteins close to their isoelectric points, and at low salt concentrations, the effects of salts have been shown to be very small [69].

The dependence of the partition coefficient at high salt concentrations in the PEG/dextran aqueous system has been explored by Albertsson [6], and results for protein partitioning in a PEG/dextran/water system with increasing NaCl concentration are presented in Fig. 13. It was found that at high NaCl concentrations proteins tend to favor the upper, PEG-rich phase. This could possibly be due to hydrophobic interactions with the PEG or salting out [6].

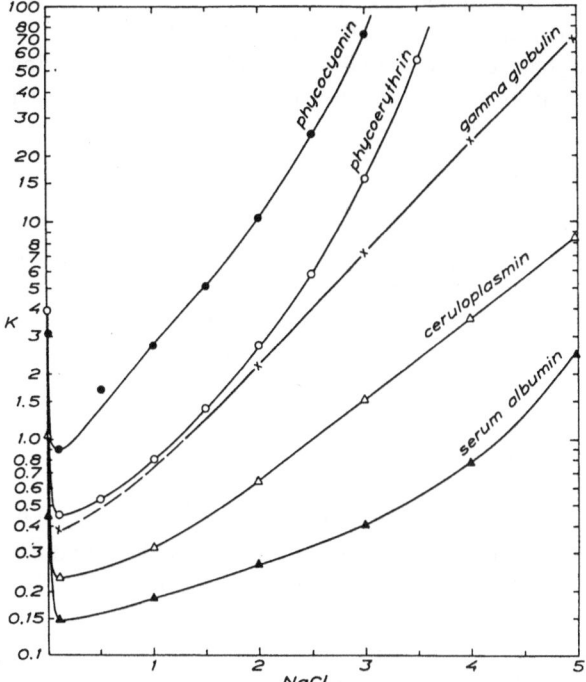

**Fig. 13.** Partition coefficient of proteins in a PEG 8000/Dextran T-500/Water system with 0.005 M $KH_2PO_4$, 0.005 M $K_2HPO_4$, and with increasing concentration of NaCl. NaCl concentration is expressed as moles NaCl added per kilogram standard phase system. From Albertsson [6]

When discussing the effect of salt on aqueous polymer phase diagram it was noted that the salts will partition unequally between the phases. This unequal distribution, along with the requirement of electroneutrality in the phases leads to an electrostatic potential difference, $\Delta\psi$, between the phases [34, 74]. For example, Reitherman et al. [75] measured $\Delta\psi$ in a PEG/Dextran system containing both sodium phosphate and sodium chloride and found that it varied from 2.8 mV to approximately 0.0 mV as sodium phosphate was varied from 0.11 to 0.011 M and KCl was increased from 0.0 to 0.126 M. Brooks et al. [76] determined $\Delta\psi$ to be 2 to 3 mV in a PEG/dextran system containing potassium sulfate concentrations from 0.001 $mol\,kg^{-1}$ to 0.4 $mol\,kg^{-1}$. King et al. [43] found $\Delta\psi$ to vary between 0 and 7 meV in PEG/dextran systems containing $K_2SO_4$, $KH_2PO_4$, KCl and NaAc.

Albertsson [7] derived the following relationship between the electrostatic potential difference and the partition coefficients of the ions in the absence of a potential ($K_-$ and $K_+$):

$$\Delta\psi = \frac{RT}{(z^+ + z^-)F} \ln\left(\frac{K_-}{K_+}\right) \tag{18}$$

Therefore, the greater the difference in affinity of a salt for the two phases, the greater the $\Delta\psi$. Diamond and Hsu [36] have also shown that the electrostatic potential in PEG/dextran systems can be related to the polymer concentration difference between the phases according to the following empirical second order relationship:

$$\Delta\psi = g(w_1'' - w_1') + h(w_1'' - w_1')^2 \tag{19}$$

where g and h are the parameters in Eq. (11) which was developed from Flory-Huggins theory for correlating protein partitioning. It was found that $\Delta\psi$ would increase in magnitude as the polymer concentration difference between the phases was increased (i.e., further away from the plait point). Earlier, Albertsson [7] developed a general relationship between the partition coefficient of a charged biomolecule and the electrostatic potential difference:

$$\ln K_3 = \ln K_{3,o} + \frac{z_b F}{RT} \Delta\psi \tag{20}$$

where $K_{3,o}$ is the partition coefficient of the biomolecule when $\Delta\psi = 0$ or $z_b = 0$. Equation (11) is a more developed form of Eq. (20) in which the $\ln K_{3,o}$ and $\Delta\psi$ have been expressed in terms of the polymer concentration difference.

## 3.5 Effect of Charged Polymer Derivatives

The partitioning of proteins can be greatly influenced by the addition of charged phase forming polymer derivatives to aqueous two-phase systems [77, 78]. Johansson [77] synthesized trimethylamino-PEG (TMA-PEG) and sulfonate-PEG (S-PEG) derivatives and used these to investigate the partitioning of bovine serum albumin, human serum albumin and ovalbumin in PEG/dextran based systems. The PEG (underivatised) was replaced by 0 to 50% TMA-PEG or S-PEG. The results of the pH investigation are presented in Fig. 14. It was found that when the net charge on the protein was the same as that for the charged PEG derivative, the protein partition coefficient decreased (i.e. more partitioning into the bottom, dextran-rich phase) the higher the concentration of the PEG derivative. However, when the protein had a charge opposite to that of the PEG-derivative, its partition coefficient increased (i.e., more partitioning into the PEG/PEG-derivative rich phase). In a later study, Johansson et al. [78] investigated the effects of pH and added salt on the protein partition coefficient in systems containing charged PEG's. The protein utilized was co-hemoglobin, which has an isoelectric pH of 6.9. It was found that in a PEG/dextran system containing TMA-PEG and at pH 6.0, the co-hemoglobin (which was positively charged) would partition to the bottom phase. As the pH was increased above 6.9, the partition coefficient also increased. The partition coefficient was shifted from approximately 0.01 to 10.0 in the pH range 6.0 to 8.0. It was also discovered that the addition of salts to the system tended to suppress the effect of the TMA-PEG. This was probably due to both a drop in the electrical potential

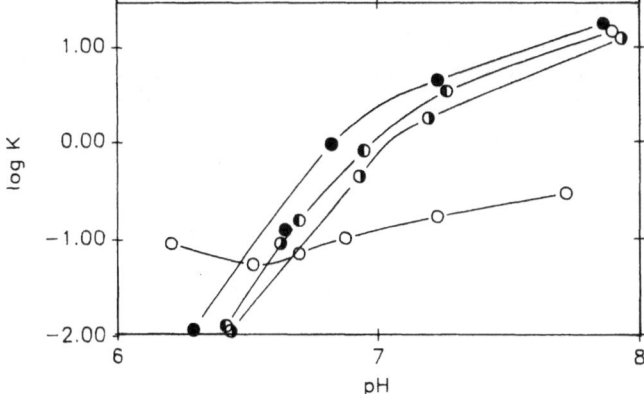

**Fig. 14.** Partition coefficient of CO-hemoglobin in the presence of charged PEG. The phase system consists of 8% Dextran T-500 and 8% PEG 8000 (I) plus Trimethylamino-PEG (II) ○, I only; ◐ 50— I and 50% II; ◑ 25% I and 25% II; ● II only. From Johansson et al. [78]

difference [6], and by salts effectively masking the electrical interaction between charged polymer and the protein charges [60]. However, zwitterions such as glycine could be added without producing any appreciable changes in the effect of the TMA-PEG [78].

Hughes and Lowe [79] used polyampholytic acrylic co-polymers together with polyvinyl alcohol (PVA) for aqueous two-phase systems. The acrylic heteropolymers did not have charge heterogeneity, and because of their poly-ampholytic properties, behaved as buffers. The two-phase systems were used to purify human serum albumin, trypsin and carboxypeptidase. Interestingly, proteins could be recovered from the liquid polyampholyte phase by isoelectric precipitation of the polymer.

Cheng et al. [80] investigated the partitioning of glucose-6-phosphate de-hydrogenase and other yeast proteins in PEG/dextran aqueous two-phase systems containing polymers with charges and affinity-dye ligands. It was found that when the polyelectrolyte was forced into one phase, it could be used to strongly influence the partitioning of the proteins. For example, the partition coefficient of glucose-6-phosphate dehydrogenase could be changed by as much as 1700-fold in a PEG/dextran system by the addition of DEAE-dextran and potassium iodide.

## 3.6 Effect of Hydrophobic Polymer Derivatives

The partition coefficient of proteins in PEG/dextran aqueous systems can be strongly influenced by the addition of hydrophobic PEG derivatives [81–86]. Two key examples are discussed below. Shanbhag and Johansson [81] syn-thesized PEG-palmitate (P-PEG) and added it to a PEG/dextran based system from purification of human serum albumin from plasma. Human serum albumin is known to bind non-polar compounds so the experimental investigation can actually be considered an affinity purification (see next section). It was demonstrated that the partition coefficient for human serum albumin in the absence of P-PEG was essentially zero (i.e. partitioning into the bottom phase). When the P-PEG concentration was increased from 0 to 10% the partition coefficient was increased to 3.15.

Shanbhag and Axelsson [82] measured the partition coefficient of a series of proteins (bovine serum albumin, β-lactoglobulin, hemoglobulin and cytochrome c) in a PEG/dextran system containing PEG bound fatty acids. The carbon chains of the covalently bound fatty acids contained between 2 and 20 carbon atoms. The partition coefficients in their investigation were reported as $\Delta \ln K$ which was defined as the natural logarithm of the partition coefficient of the proteins in the presence of the PEG-fatty acid minus the natural logarithm of the partition coefficient in its absence. It was generally observed that the $\Delta \ln K$ was greatest for each of the above proteins when the fatty acid chain length was longer than eight carbon atoms.

## 3.7 Effect of Affinity Polymer Derivatives

One of the most promising areas for application of aqueous two-phase systems is the affinity purification of biomolecules through attachment of affinity ligands to the phase forming polymers. The purification of a protein by affinity partition is based on the preferential (or biospecific) interaction between the biomolecule and the affinity polymer derivative, and results in a biomolecule-polymer derivative complex which partitions to one phase while the contaminating proteins partition predominantly to the opposite phase. Affinity partitioning has been primarily investigated in PEG/dextran aqueous two-phase systems containing affinity PEG derivatives in which conditions are sought whereby the contaminating proteins partition to the lower, dextran-rich phase while the target biomolecule is selectively partitioned into the upper, PEG-rich phase with the help of the affinity PEG derivative [87]. PEG/salt systems have been used rather effectively for the affinity purification of human fibroblast interferon [88] in which a phosphate derivative of PEG was used as the affinity ligand. However, this appears to be one of the few affinity two-phase investigations using the PEG/salt due to the high salt concentrations required for its formation. The high salt concentrations pose a potential interference with the biospecific interactions.

In Table 2 a summary of protein purifications using affinity aqueous two-phase systems is presented, and some of the pioneering results are discussed below. One of the first attempts at biospecific affinity partitioning was the hydrophobic partitioning work of Shanbhag and Johansson [81] which was discussed in the preceding section. They had used a PEG-palmitate derivative to

**Table 2.** Affinity partitioning of biomolecules in aqueous two-phase systems

| Biomolecule | Ligand | Reference |
|---|---|---|
| Trypsin | Diamino-$\alpha$, $\omega$-Diphenyl Carbamyl | [89] |
| Trypsin | Trypsin Inhibitor | [90] |
| Serum Albumin | Fatty Acid | [81, 83, 85] |
| $\beta$-Lactoglobulin | Fatty Acid | [82] |
| Concanavalin | — | [91] |
| S-23 Myeloma Protein | Dinitrophenyl | [91] |
| Histones | Fatty Acid | [92] |
| 3-Oxosteroid Isomerase | Estradiol | [93] |
| Formaldehyde Dehydrogenase | NADH | [94] |
| Formate Dehydrogenase | NADH | [94] |
| Formate Dehydrogenase | Procion Red | [95] |
| Colipase | Lecithin | [96] |
| Myosin | Fatty Acid | [97] |
| Phosphofructokinase | Triazine Dye | [98–102] |
| Interferon | Phosphate | [88] |
| Pyruvate Kinase | Triazine Dye | [101] |
| Glutamate Dehydrogenase | Triazine Dye | [101] |
| Glycerol Kinase | Triazine Dye | [101] |
| Hexokinase | Triazine Dye | [100, 101, 103] |
| Lactate Dehydrogenase | Triazine Dye | [101, 102, 104–108] |
| Malate Dehydrogenase | Triazine Dye | [101] |
| Transaminase | Triazine Dye | [101] |
| $\alpha$-Fetoprotein | Triazine Dye | [109] |
| Pre-Albumin | Remazol Yellow | [109, 110] |
| Glucose-6-Phosphate Dehydrogenase | Triazine Dye | [62, 100, 103, 108, 111–113] |
| Glucose-6-Phosphate Dehydrogenase | Triazine Dye and Charged Groups (DEAE, Sulfate) | [80] |
| Glyceraldehyde Phosphate Dehydrogenase | Triazine Dye | [62, 100] |
| 3-Phosphoglycerate Kinase | Triazine Dye | [62, 100, 108] |
| 3-Phosphoglycerate Kinase | ATP | [114] |
| Alcohol Dehydrogenase | Triazine Dye | [62, 100] |
| Nitrate Reductase | Triazine Dye | [115] |
| Acid Proteases | Pepstatin | [116] |
| Thaumatin | Glutathione | [90] |
| IgG | Protein A | [90] |
| Human Hemoglobin | Cu(II)IDA | [117] |
| Vancomycin | D-ala-ala-ala | [118] |
| Cytochrome c | Cu(II)IDA | [119] |
| Myoglobins | Cu(II)IDA | [119] |
| Hemoglobins | Cu(II)IDA | [119] |
| $\alpha_2$-Macroglobulin | Metal-IDA | [120] |
| Tissue Plasminogen Activator | Metal-IDA | [120] |
| Superoxide Dismutase | Metal-IDA | [120] |
| Monoclonal Antibodies | Metal-IDA | [120] |
| Membranes from Calf Brain Synaptosomes | Procion Yellow HE-3 | [121] |
| Albumins | Alcohols | [122] |
| Thylakoid Membranes | Alcohols | [122] |

specifically attract human serum albumin into the upper phase of a PEG/ dextran system. Takerkart et al. [89] utilized *p*-aminobenzamidine, which is a trypsin inhibitor, as a PEG affinity ligand in a PEG/dextran system for the biospecific purification of trypsin. The term "affinity partitioning" was first coined by Flanagan and Barondes [91] who used an affinity PEG/dextran system for the purification of S-23 myeloma protein. The affinity system contained dinitrophenyl-PEG, which specifically binds to the S-23 myeloma protein, to the system. The partition coefficient of the S-23 myeloma protein was increased from 2.8 in the absence of the dinitrophenyl-PEG to 7.0 in the presence of 0.18 mM dinitrophenyl-PEG. Hubert et al. [93] successfully purified $\Delta_{5 \to 4}$ oxosteroid isomerase using a PEG estradiol derivative. It was shown that by introduction of the PEG-derivative, the isomerase partition coefficient could be increased three-fold (from 5 to 15) while the bulk protein remained in the bottom phase. Kopperschläger and Johansson [98] pioneered the use of PEG-affinity dye-ligands for protein purification. Utilizing Cibacron blue F3G-A-PEG, they affinity purified phosphofructokinase (PFK) from baker's yeast. A 58-fold purification of the PFK was achieved within three hours.

In order to more effectively and efficiently utilize affinity aqueous two-phase systems, a need exists for a fundamental understanding of the affinity partition phenomenon. A number of parameters have been shown to influence the affinity partition of proteins and these include ligand concentration and binding characteristics, polymer concentration, salt concentration, molecular weight of polymers, pH, temperature, number of ligands per molecule, organic solvents, and bulk protein [123]. A theory which is able to predict the effects of each of the above parameters has not yet been developed. However, affinity partition theories have been developed which focus on the effects of ligand concentration and binding and number of ligands per molecule [91, 124, 125].

One of the first attempts at the mathematical modeling of affinity protein partition was that by Flanagan and Barondes [91]. Their experimental work with S-23 myeloma protein is discussed above. For proteins with n independent binding sites and systems with very high ligand concentrations the following partition expression was developed [91]:

$$K_3 = K_{L,o} \left( K_L \frac{K_a''}{K_a'} \right)^n \tag{21}$$

where $K_3$ is the biomolecule partition coefficient in the presence of polymer ligand, $K_{L,o}$ is the biomolecule partition coefficient in the absence of polymer ligand and can be determined by Eq. (11), $K_L$ is the partition coefficient of the polymer bound ligand, and $K_a''$ and $K_a'$ are the binding constants for the protein-ligand complex in the top and bottom phases, respectively. The limitation of this relationship is that it is only useful in the limit of high ligand concentration. Cordes et al. [124] developed a more general relationship for Eq. (21) that can be applied to affinity systems with any ligand concentration:

$$K = K_{L,o}\left(\frac{1 + L''K_a''}{1 + L''K_a'/K_L}\right)^n \tag{22}$$

where $L''$ is the ligand concentration in the top phase. Their experimental work focused on the partitioning of the enzyme formate dehydrogenase (FDH) in the presence of triazine dyes or NADH attached to PEG as the affinity ligands. FDH has two binding sites for NADH, and thus, it was found that by modifying Eq. (22) to include different binding constants for two ligand binding sites in the PEG-rich phase and one site in the dextran-rich phase, the partition data could be more adequately represented. Based on their experimental and theoretical work, it was concluded that the greater the binding between the desired enzyme (i.e. the enzyme to be purified) and ligand, the lower the concentration of ligand required to be added to the system. Increasing ligand concentration would only result in the increased partitioning of contaminating proteins into the upper phase according to the magnitude of their dissociation constants.

One other development in affinity aqueous two-phase theory includes the modeling of metal-affinity protein partitioning in Cu(II)PEG/dextran based systems [125]. The model takes into account the complex formed between the Cu(II)PEG ligand and surface histidine residues of proteins as well as inhibition of the binding due to the presence of hydrogen ions. It is suggested that the model can be applied to other affinity two-phase systems which experience inhibition by free affinity ligands or inhibitors.

A very promising application of affinity aqueous two-phase systems is the large-scale purification of enzymes from a variety of sources. To accomplish this one must consider the choice of polymer for the two-phase system, the affinity ligand and the technical and economical feasibility of the overall extraction process. Several groups have evaluated the large-scale affinity two-phase purification of enzymes [95, 105]. Cordes and Kula [95] purified formate dehydrogenase (FDH) with affinity partitioning directly from a cell homogenate of *Candida boidinii*. The affinity ligand was Procion Red HE3b attached to PEG 8000. The process consisted of first subjecting a 40% cell suspension (44 kg of cells) to heat denaturation. PEG, crude dextran and red-PEG were then added to the product from the heat denaturation step resulting in an affinity two-phase system consisting of 9% PEG, 1% crude dextran and 1 mmol $l^{-1}$ of the affinity ligand. The FDH was specifically attracted by the red-PEG into the upper, PEG-rich phase while the contaminating proteins were partitioned to the lower, crude dextran-rich phase. The phases were separated, and potassium phosphate was then added (up to a concentration of 9%) to the top phase containing the FDH and red-PEG. This resulted in a PEG/potassium phosphate two phase system which was to be used to separate the FDH from the red-PEG. In this system, the FDH partitioned to the lower, salt-rich phase, while the red-PEG partitioned to the upper, PEG-rich phase. The phases were separated, and the upper phase containing the red-PEG was recycled. Fresh PEG was added to the lower phase (containing the FDH), and a new PEG/potassium phosphate system resulted. The purpose of this step was to further remove the red-PEG from the FDH. The phases were again separated, and the top phase containing the red-PEG was recycled. The lower, salt-rich phase containing the FDH was then subject to ultrafiltration and lyophilization resulting in the final product.

A second example of a large-scale affinity two-phase process was the purification of lactate dehydrogenase (LDH) from pig muscle using a PEG/hydroxypropyl starch two-phase system with Procion Yellow HE-3G coupled to PEG as the affinity ligand [105]. The process was begun by either 1) homogenizing the muscle followed by removal of debris by centrifugation and then introduction into the PEG/hydroxypropyl starch affinity two-phase system, or 2) direct homogenization in the affinity two-phase system. Either way, the lactate dehydrogenase was selectively attracted into the upper PEG-rich phase by the yellow-PEG. The phases were separated, and yellow-PEG was then removed from the LDH by formation of a PEG/potassium phosphate system. The LDH was partitioned to the lower, potassium phosphate-rich phase and the yellow-PEG was partitioned to the upper, PEG-rich phase. The upper phase was recycled and the LDH was subsequently recovered from the lower phase. Economic analysis showed that the affinity aqueous two-phase process could purify the LDH for US\$ $0.13\,\mathrm{kU}^{-1}$ or US\$ $0.23\,\mathrm{kU}^{-1}$ depending on whether the process was started with procedure (1) or (2), respectively. These costs were approximately 1/10 and 1/5 the current selling price of LDH (US\$ $1.3\,\mathrm{kU}^{-1}$ in 1987), respectively.

## 4 Large-Scale and Multi-Stage Purification of Proteins

The large-scale purification of proteins using aqueous two-phase systems provides an attractive alternative (both technically and economically) to traditional bioseparations processes [27, 126, 127]. The extent to which these systems have been applied or considered for large-scale downstream processing of proteins can be effectively divided into two categories: 1) a few (one to three) mixing/settling stages to achieve purification or partial purification of intracellular microbial enzymes, and 2) multi-stage processing using countercurrent distribution (CCD), liquid-liquid partition chromatography (LLPC), or continuous countercurrent chromatography (CCC) to improve resolution of the protein products. The first category has been more extensively researched than the second, and a summary of enzyme purification using a few extraction stages is presented in Table 3. The extraction procedure of category 1 has been demonstrated to be suitable for the early stages of enzyme purification, i.e., the removal of cell debris, but may also be used to partly replace chromatographic steps. The second category has primarily been investigated on a laboratory-scale and has significant potential for large-scale applications including purification of high purity protein product. Each of the categories will be discussed below.

**Table 3.** Large-scale biomolecule purification using aqueous two-phase systems

| Biomolecule | Reference |
|---|---|
| Acyl Aryl Amidase | [127] |
| Alcohol Dehydrogenase | [127] |
| α-Amylase | [128] |
| Aspartase | [127, 129] |
| Aspartate β-Decarboxylase | [127] |
| Chlorophyll a/b-Protein (LHPC) | [130] |
| Chromatophores | [131] |
| Formaldehyde Dehydrogenase | [26] |
| Formate Dehydrogenase | [26, 127, 132] |
| Fumarase | [27, 127] |
| β-Galactosidase | [133, 134] |
| Glucose Dehydrogenase | [127] |
| Glucose Isomerase | [135] |
| Glucose-6-Phosphate Dehydrogenase | [127, 136] |
| α-Glucosidase | [127] |
| Hexakinase | [127] |
| D-2-Hydroxyisocaproate Dehydrogenase | [127, 137] |
| L-2-Hydroxyisocaproate Dehydrogenase | [138] |
| Interferon | [88] |
| Isoleucyl-tRNA Synthetase | [126] |
| Isopropanol Dehydrogenase | [127] |
| D-Lactate Dehydrogenase | [139] |
| Leucine Dehydrogenase | [127, 140] |
| NAD-Kinase | [127] |
| Pencillin Acylase | [127] |
| Phosphofructokinase | [98, 136] |
| Phospholipase | [141] |
| Phosphorylase | [25] |
| Pullulanase | [25, 127] |
| *Staphylococcal* Protein A-β-Galactosidase Hybrid | [142] |
| Whey Proteins | [143] |

## 4.1 Large Scale Extraction Using a Few Stages

One of the key factors in the large-scale purification of proteins using aqueous two-phase systems is the selection of an appropriate system. For the most part, PEG/dextran and PEG/salt have been used for the majority of the purifications reported in the literature. These systems appear to possess many desirable characteristics including a general applicability; the physico-chemical properties of the phases (viscosity, density difference) are relatively suitable; and the polymers are nontoxic, biodegradable, and certified in the Pharmacopoeias of most countries [127]. The high cost of fractionated dextran (approximately US$ 500 per kg) in the PEG/dextran systems and the high concentration of salt in the PEG/salt systems have led researchers to investigate suitable alternatives. Kroner et al. [26] evaluated the PEG/crude dextran system for large-scale enzyme purification and found no serious technical problems with its use. In addition, crude dextran sells for approximately US$ 15 per kg which compares rather favorably with the fractionated type. However, a major drawback of the PEG/crude dextran systems is the high viscosity of the lower phase, although this was reduced by partial hydrolysis of the crude dextran [26]. Tjerneld et al. [9] introduced hydroxypropyl starch for replacement of dextran. The PEG/ hydroxypropyl starch system was shown to have similar characteristics to the PEG/dextran systems and the hydroxypropyl starch sells for approximately US$ 20 per kg. The successful large-scale purification of pig muscle using an affinity PEG/hydroxypropyl starch system [105] was discussed in the previous section.

Other inexpensive phase forming polymers which have been considered include ethylhydroxyethyl cellulose (EHEC) [14] and maltodextrins [10, 144]. The EHEC forms two phases with hydroxypropyl starch or dextran and it is possible to obtain systems with low polymer concentrations (1% to 2% total polymer) [14]. However, the high viscosity of aqueous EHEC solutions and the long settling time of its two phases limits its usefulness in downstream processing of biomolecules. The maltodextrins for two phases with PEG are thus, a possible alternative to dextran. However, the PEG/maltodextrin systems must be more fully investigated to determine their usefulness.

Hsu [15] has developed novel aqueous two-phase systems which appear to be very promising for large-scale biochemical separations. These systems are based on the ability of PEG or polypropylene glycol (PPG) to form two phases when mixed with an amino acid, protein, peptide, monosaccharide, disaccharide, or chiral compound. These systems are capable of resolving mixtures of optical isomers such as other amino acids. The novel two-phase systems are also capable of purifying proteins, cell particulates, nucleic acids, etc.

After selecting the system to be used, the next step involves selection of the appropriate system conditions. When using a few extraction steps for removal of cell debris and subsequent partial purification of the desired protein, the goal is to get the cell debris and desired protein to partition to opposite phases. The factors to be considered when selecting system conditions include the partition coefficients of the target protein and contaminating materials, the volume ratio of the phases and the concentration range of broken cells [127].

When the protein is to be purified into the top phase, system conditions should be adjusted such that its partition coefficient is as high as possible while that of the cell debris and contaminating proteins are as low as possible. In order to achieve this, the basic experimental and theoretical principles developed in Sects. 1 and 2 can be applied. For example molecular weight, pH, the addition of affinity ligands, etc.

With the assumption that the desired protein is to be extracted into the top phase, Hustedt et al. [127] have presented an equation for the theoretical yield of the target enzyme in the top phase as a function of the volume ratio of the phases $(V''/V')$ and the protein partition coefficient, $K_3$

$$Y_3'' = \frac{100}{1 + (V''/V')(1/K_3)} \, (\%) \tag{23}$$

where $Y_3''$ is the yield of the target protein in the top phase and $V''$ and $V'$ are the top and bottom phase volumes, respectively. Generally, as the ratio of top to bottom phase is increased at constant $K_3$, the protein yield in the top phase is also increased. Similarly, as $K_3$ is increased at constant phase volume ratio, the yield in the top phase will also increase. Volume ratios between 3 and 4 are most readily used in practice [127] and the enzyme partition coefficients are mainly in the range of 3 to 20 [127], thus, giving an enzyme yield of approximately 90% in the top phase.

When selecting two-phase system conditions, one other parameter, the broken cell concentration, should be considered. Cell debris will contribute to the formation of the phases and will generally decrease the phase volume ratio as its concentration is increased [127]. The concentration of broken cells that has often been used is between 20 to 30%, but has not yet been optimized [127].

One of the final considerations in the use of aqueous two-phase systems for large-scale enzyme purification is separation of the target protein from the phase forming polymers, in particular, PEG. Methods which have been investigated include first transferring the target protein into a relatively polymer-free phase such as the lower, potassium phosphate-rich phase of the PEG/potassium phosphate system [127]. The concentration of PEG is normally low in this phase ( < 2%) and the PEG and salts may subsequently be removed by ultrafiltration and desalting [127]. PEG and desalting can also be accomplished by diafiltration followed by ultrafiltration for concentration of the protein product [137]. PEG may also be removed by chromatographic adsorption on hydroxyapatite or ion-exchange based columns [6, 127].

Albertsson [6] has described a method which is applicable to proteins in PEG/dextran systems. If the protein partitions to the lower phase, then ammonium sulfate or potassium phosphate may be added to the system. As salt is added, the PEG concentration in the lower phase diminishes to the point where it is completely removed from the lower phase. The protein in the lower phase may then be precipitated by further additions of ammonium sulfate.

A general protein purification scheme for a one to three step aqueous two-phase process is given in Fig. 15. The first step entails the partitioning of disrupted cells in a PEG/salt or PEG polymer system. The cell debris will generally partition to the bottom phase while system conditions can be adjusted such that the target protein partitions to the top phase. If the purpose is this step is solely for removal of cell debris, then the two-phase processing may stop here [127] and the purification will subsequently continue with concentration and high resolution steps. Alternatively, this single step may provide the desired protein purity and the enzyme can then be removed from the polymers or salt using the methods described above. Still another alternative is to continue with one to two additional aqueous two-phase extraction steps to achieve greater protein purity [127]. Examples of protein extraction using aqueous two-phase systems will be discussed below.

A single step extraction process was used by Veide et al. [133] for the large-scale isolation of β-galactosidase from a suspension of disintegrated E. coli cells.

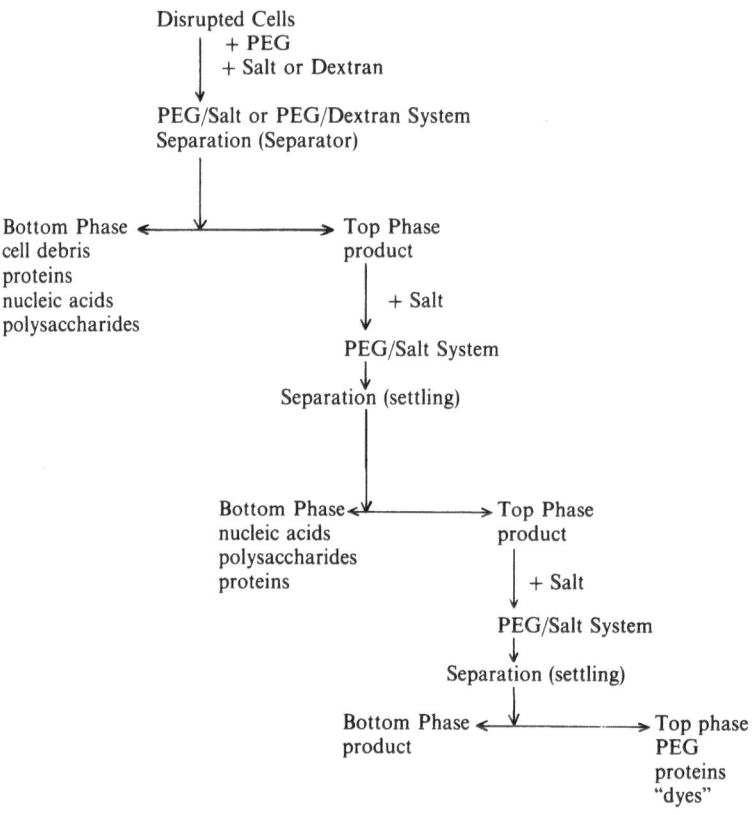

**Fig. 15.** General protein purification scheme using one to three extraction steps. From Hustedt et al. [129]

The process started with a 372 l *E. coli* culture which contained 7.5 U mg$^{-1}$ and a total of $26.5 \times 10^6$ units of β-galactosidase. The cells were concentrated by centrifugation and then disintegrated with a glass bead mill. 47 l of disintegrated cell suspension was mixed with PEG 8000 and potassium phosphate to form a two-phase system with 6.3% PEG 8000, 10% potassium phosphate and 11% cells (wet weight). The phases were separated with a tubular bowl centrifuge. The bottom phase contained the cell debris and contaminating proteins, and the top phase contained the β-galactosidase at 75% yield ($20 \times 10^6$ units, 260 U mg$^{-1}$). The main reason why only a single extraction step produced such effective results was the rather high partition coefficient of β-galactosidase ($K = 62$) in the PEG/potassium phosphate system.

Hummel et al. [139] performed a large-scale extraction of D-lactate dehydrogenase from *Lactobacillus confusus* using 2 partition steps followed by ultrafiltration and DEAE-cellulose chromatography for subsequent concentration. 24 kg of cells (initially frozen) were disrupted by two passes through a bead mill.

The cell debris was then removed by partition in a PEG 1500/potassium phosphate system. This corresponds to the first step in Fig. 15. The lower phase contained the cell debris and the enzyme partitioned to the upper phase (K = 4.8). A second partition step was then used for further purification of the enzyme. PEG, potassium phosphate, sodium chloride and water were added to the top phase from the first step to produce the two-phase system. The D-lactate dehydrogenase was partitioned into the lower, salt-rich phase. After this second step, a crude enzyme free of cell debris was obtained with a specific activity of approximately 5.4 U mg$^{-1}$ with phenylpyruvate as substrate. The lower phase was further processed using ultrafiltration for concentration and DEAE-cellulose chromatography for further purification. These two steps resulted in a threefold enrichment of the D-lactate dehydrogenase.

## 4.2 Multi-Stage Extractions

Multi-stage liquid extraction with aqueous two-phase systems has been explored using several techniques including countercurrent distribution (CCD), liquid-liquid partition chromatography (LLPC) and continuous countercurrent chromatography (CCC). These techniques have the potential to resolve complex mixtures of proteins where a single or three step partition process is inadequate. Each of the techniques is discussed below.

### 4.2.1 Countercurrent Distribution

CCD was originally designed for aqueous-organic and organic-organic two-phase systems [4, 145]. In theory, it is very similar to column chromatography. In practice, a CCD apparatus simulates the continuous countercurrent movement of the top and bottom phases using a large number of mixing-setting chambers. A typical CCD unit for aqueous two-phase extraction consists of two circular plates with 60 or 120 cavities [6, 146]. The bottom plate is stationary while the top plate can rotate. When the plates are fit together, the cavities form 60 or 120 closed chambers. The operation of the CCD consists of shaking the two phases in a chamber, then permitting them to settle with gravity [147, 148] or centrifugation [149]. The top plate is then rotated in a clockwise manner such that the top cavity is lined up with the next cavity in the bottom plate. The top phase will contact a fresh bottom phase and vice versa. In this way, a large number of extraction steps (60 to 120) can be used to separate substances with different partition coefficients. CCD with aqueous two-phase systems has effectively been used on a laboratory scale for fractionation of proteins, cells and cell organelles [6, 17]. However, laboratory CCD devices are not available commercially, thus limiting the widespread use of the technique.

*4.2.2 Liquid-Liquid Partition Chromatography*

LLPC with aqueous two-phase systems has generally taken three modes of design and operation. The first mode was pioneered by Albertsson [7] and Blomquist and Albertsson [150] for the purification of proteins, nucleic acids and particles, and is in the form of a vertical column containing alternating mixing and settling chambers. These columns are a modified form of the industrial liquid-liquid extraction columns described by Treybal [151] and Craig [4], and have been used for aqueous-organic and organic-organic phase systems. Generally, the two phases move in opposite directions through the column with alternating mixing and separation. Alternatively, one of the liquid phases may be stationary in the column while the other is mobile. The columns are apparently easy to use experimentally, however high viscosities of the PEG/dextran phases makes the flow through the column rather slow (0.1 to 0.15 ml min$^{-1}$) [145]. Husted et al. [152] and Kula et al. [153] reported a much higher feed rate of 8 to 10 ml min$^{-1}$ using a PEG/salt system (which generally has lower phase viscosities than the PEG/dextran system) for the purification of formate dehydrogenase form *Candida boidinii*. They were able to process 6 to 24 g of protein per day using a column with a 2 cm diameter and 200 ml volume. Their stage efficiency was between 50% and 60% and overall enzyme yield was 89% compared to a theoretical value of 92%. The limited data available on the use of LLPC columns with aqueous two-phase systems suggests that their future use in this field will require systematic and detailed studies for improvement of column design and performance.

In the second mode of LLPC operation, one phase is adsorbed on an inert support, while the second phase is mobile. This form of liquid-liquid partition chromatography was pioneered by Martin and Synge [154] who used organic-aqueous systems consisting of chloroform and water for separation of amino acids from a silk hydrolysate. The water phase was bound to a silica support while the chloroform (organic) phase was mobile. Morris [155] performed an LLPC with a PEG/dextran system in which the dextran phase was adsorbed on celite, a purified kieselguhr, and synthetic calcium silicates, while the PEG phase was mobile. There was difficulty in separating the proteins (lysozyme, ribonuclease, cytochrome c, ovalbumin and bovine serum albumin) which had partition coefficients on the order of 0.1 to 0.8 despite obtaining well defined symmetrical distributions. A suggested reason for this problem was the low volume of stationary phase bound to the support per unit bed volume [156].

Müller et al. [157] later discovered that cellulose binds the dextran phase of the PEG/dextran system. Müller and Kutemeier [158] were able to efficiently separate DNA fragments of 150 to 21 000 base pairs using the dextran-phase coated cellulose support. However, the major drawbacks of using the cellulose support included its non-specific binding of single-stranded nucleic acids and proteins, and the fact that the bulk of the dextran phase was found to be hidden within the cellulose fibers making it inaccessible for small biopolymers, i.e. 20 000 to 30 000 MW [156]. It was also discovered that polyacrylamide gels, like cellulose, bind the dextran phase of the PEG/dextran system [158]. However, it was also found that rigid polyacrylamide and agarose-polyacrylamide gels tended to hide the dextran making it accessible to only small molecules (i.e. < 20 000 MW), while gels with larger pore sizes tended to swell, resulting in elastic spheres which limited flow rates [159].

The above results led Müller to graft polyacrylamide chains to standard agarose and silica based supports [160]. These new supports were found to have a much higher capacity to bind the dextran phase and better performance than cellulose supports. They were shown to be effective for the fractionation of DNA restriction fragments ranging from 11 base pairs to 3829 base pairs. It has also been used for the separation of proteins [156, 161, 162].

The third mode of LLPC involves the covalent attachment of one of the phase forming polymers to a solid chromatographic support, while the other polymer phase (and a possible addition of a small amount of the polymer bound phase) are passed through the column as the mobile phase. The technique has been explored by Matsumoto and Shibusawa [163] who covalently attached PEG to silica beads and Sepharose 6B, and then used dextran in the mobile phase buffers. The columns have been used to successfully separate blood cells [163], and later papers reported the effects of mobile phase composition [164], molecular weight of the PEG bound [165], surface hydrophobicity of the blood cells [166], and the use of polypropylene glycol (PPG) in place of PEG [167, 168], on elution of the cells from the column. It should be pointed out that PPG forms two phases with dextran, and was recorded in Table 1.

### 4.2.3 Continuous Countercurrent Chromatography

Continuous countercurrent chromatography evolved from the observation that two immiscible liquids flowing countercurrently in a helical tube, which is rotating in an acceleration field, become uniformly segmented in the coils of the tube [169]. CCC does not have discrete transfer steps as in CCD or LLPC (without a solid support), rather its continuous operation provides it with the capability of achieving a high theoretical number of transfers as exhibited in HPLC. Since there is no solid support with CCC, it is free from all complications arising from them. Adsorptive loss, biomolecule denaturation and sample contamination are minimized. When aqueous two-phase systems are used with CCC there are the added advantages of aqueous two phase processing discussed in Sect. 1, one of the most important being that the phase forming polymers tend to stabilize the labile biomolecules.

Two CCC designs (or modifications thereof) have been used with aqueous two-phase systems. The toroidal coil consists of PTFE tubing helically wound around the circumference of a rotating plate [170]. If a PEG/dextran system is used, then the more dense dextran-rich phase is maintained in the tubing by centrifugal force, while the upper PEG-rich phase is mobile. The toroidal coil has been used for fractionation of bacterial cells [170], subcellular fractionation of rat liver homogenates [171] and with affinity partition [172].

The second type of CCC is the coil planet centrifuge. This CCC, which has several functional designs, is described below. The operation of the device can be equated to the rotation of a planet about its axis and its revolution about the sun. A column holder (i.e. the planet) can have PTFE tubing wrapped around it in a multilayer, eccentric, or toroidal form [173]. The column rotation about its axis can be parallel [174] or perpendicular [175] to the axis of revolution. The axis or revolution is a stationary shaft (the sun). When the axis of rotation is perpendicular to the shaft the coil planet centrifuge has been referred as being "cross-axis" [175]. The coil planet centrifuge may also be designed as synchronous or nonsynchronous depending on the rotation of the column with respect to its revolution [176]. The device is synchronous if the "planetary" gear interlocks the "sun" gear which is counted around the stationary central shaft. The rotation and revolution are, therefore, driven by a single motor. In the nonsynchronous mode, two motors are utilized, one to generate rotation of the columns and the other to generate revolution about the stationary shaft.

Aqueous two-phase systems have been successfully used with a nonsynchronous coil planet centrifuge for the separation of erythocytes [176] and *Salmonella tryphimurium* [177, 178]. Recently, an eccentric, multilayer, synchronous coil planet centrifuge has been evaluated for protein separation using a PEG/potassium phosphate system [179]. BSA and lysozyme were used as a standard mixture. It was found that an optimal flow rate and revolution (rotation) speed could be obtained for separation of the proteins, but that other parameters, including coil diameter and centrifuge volume need to be further investigated to improve separation efficiency. A cross-axis synchronous coil planet centrifuge has also been successfully used for separation of proteins using a PEG/potassium phosphate system. Mixtures of cytochrome c, myoglobin, ovalbumin and hemoglobin have been separated as well as recombinant uridine phosphorylase from an crude *E. coli* homogenate [180, 181].

# 5 Extractive Bioconversions

Aqueous two-phase systems provide an interesting means for the simultaneous production and purification of a bioproduct using enzymes or microorganisms. An aqueous two-phase extractive bioconversion process is schematically illustrated in Fig. 16. Just as with large-scale protein purification or affinity partition which attempt to isolate the target protein and contaminating materials in opposite phases, the goal here is to utilize a two-phase system in which the substrate or the biocatalyst (enzyme or microorganism) partition to one phase, and the product (protein, steroid sugar, etc. . .) partitions to the opposite phase. The purpose of separating the product from the substrate is not only for purification purposes but can be used to increase the velocity of the reaction or

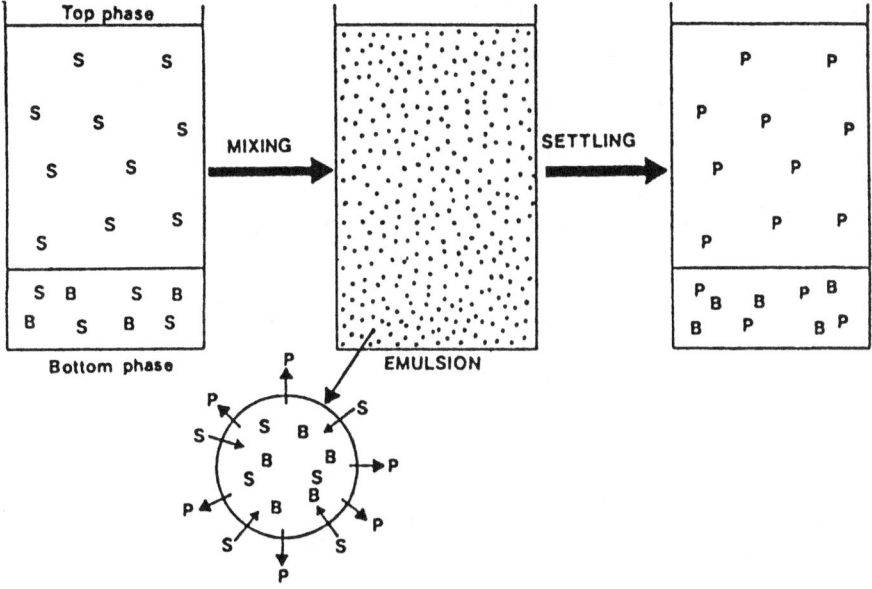

**Fig. 16.** Schematic illustration of an extractive bioconversion using an aqueous two-phase system. $S$ = Substrate, $P$ = Product, $B$ = Biocatalyst. Andersson and Hahn-Hägerdal [182]

conversion if the enzyme or microorganism is inhibited by the product [6]. Similarly, if the biocatalyst is inhibited by the substrate, conditions can be obtained in which the substrate and biocatalyst partition to opposite phases [183]. A summary of bioconversions using enzymes and microorganisms as biocatalysts is given in Tables 4 and 5, respectively. Some examples of each are discussed below.

Tjerneld et al. [186, 187] investigated the semi-continuous hydrolysis of cellulose in PEG/crude dextran two-phase systems using endo-β-gluconase and

**Table 4.** Extractive bioconversions using enzyme catalysts in aqueous two-phase systems

| Enzyme | Product | Phase System | Reference |
|---|---|---|---|
| α-Amylase and glucoamylase | Reducing Sugars | PEG/Crude Dextran | [184] |
| Acylase | L-Methionine | PEG/Phosphate | [185] |
| endo-β-Glucanase and β-Glucosidase | Reducing Sugars | PEG/Crude Dextran | [186–188] |
| β-Galactosidase | Glucose | PEG/Pullulan | [189] |
| Glucoamylase | Glucose | PEG/Starch | [190] |
| Glucoamylase and α-Amylase | Glucose | PEG/Crude Dextran | [191] |
| Glucoamylase and α-Amylase | Glucose | PEG/Dextran | [192] |
| Hexokinase and Acetate Kinase | Glucose-6-Phosphate | PEG/Dextran with PEG-Sulfate added | [193] |
| Penicillan Acylase | 6-APA | PEG/Phosphate | [194] |

**Table 5.** Extractive bioconversions using microorganisms as catalysts in aqueous two-phase systems

| Microorganism | Product | Phase System | Reference |
|---|---|---|---|
| *Arthrobacter simplex* | Prednisilone | PEG/Dextran | [195] |
| *Aspergillus phoenicis* | β-Glucosidase | PEG/Dextran PEG/PVA | [196] |
| *B. amyloliquefaciens* | α-Amylase | PEG/Dextran | [197] |
| *Bacillus licheniformis* | Alkaline Protease | PEG/Dextran | [198] |
| *B. subtilis* | α-Amylase | PEG/Dextran | [199–201] |
| *Brevibacterium ammoniagenes* | L-Malate | PEG/Phosphate | [202] |
| *Brevibacterium CH1* | Acrylamide | PEG/Phosphate | [183] |
| *Chlostridium acetobutylicum* | Acetone/Butanol | PEG/Dextran | [203] |
| *Chlostridium tetani* | Toxin | PEG/Dextran | [204] |
| *Mycobacterium* | Andros-4-ene-3,17-dione (AD), androsta-1,4-diene-3,17-dione (ADD) | PEG/Dextran | [205] |
| *S. cerevisiae* | Ethanol | PEG/Dextran | [206, 207] |
| *S. cerevisiae* | Ethanol | PEG/Crude Dextran | [208, 209] |
| *T. reesei* | Cellulase | PEG/Pullulan | [189] |
| *T. reesei* | Cellulase | PEG/Dextran | [210] |

β-glucosidase from *Trichoderma reesei*, QM9414. The experimental design for their process is given in Fig. 17. The top and bottom phases were mixed in a mixer unit, and solid cellulose was added intermittently. The mixture was pumped to a settler unit where the two phases were separated. After separation, both phases were recycled to the mixer. The top phase could also be removed for subsequent purification of glucose. The semicontinuous process was run for more than 450 h. The starting concentration of cellulose substrate was 75 g l$^{-1}$, and, with intermittent addition of the substrate, 50 g l$^{-1}$ of glucose could be produced at dilution rates ranging from 0.006 to 0.12 h$^{-1}$.

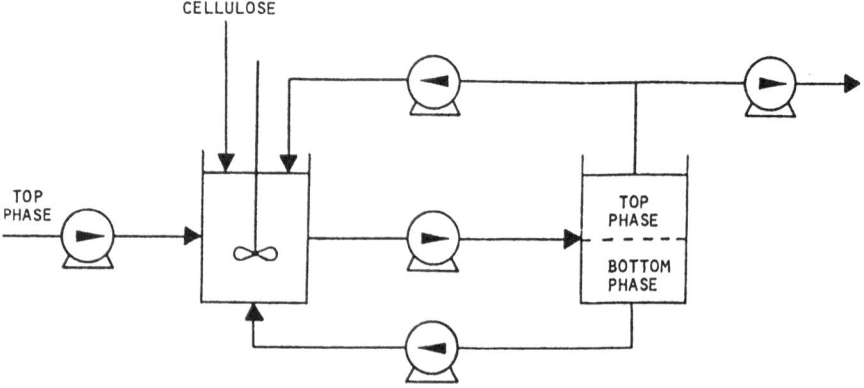

**Fig. 17.** Experimental design for semicontinuous hydrolysis of cellulose in an aqueous two-phase system. From Tjerneld et al. [187]

Another extractive bioconversion involved the continuous enzymic hydrolysis of starch in a PEG/crude dextran system using α-amylase and glucoamylase [191]. The experimental apparatus, which combined a mixing-setting tank with an ultrafiltration unit, is provided in Fig. 18. A starch suspension was fed to the mixing chamber and the top and bottom phases were recycled from the settling chamber. The top phase in the right chamber was passed through an ultrafiltration unit, yielding a continuous stream of glucose product. The process was operated continuously for eight days.

Puziss and Hedén [204] pioneered the use of microbial based bioconversions in aqueous two-phase systems. They utilized a PEG/dextran system for growth of *Clostridium tetani* and subsequent production and purification of toxin. The bacterial cells were partitioned into the lower, dextran-rich phase, while the toxin primarily partitioned into the upper, PEG-rich phase.

Kühn [206] produced ethanol in a PEG/dextran system with baker's yeast. The baker's yeast was added to a suspension of PEG, dextran, glucose and malt. After the process was completed, the phases were separated. Approximately 90% of the ethanol was partitioned to the upper, PEG-rich phase, while the remaining 10% was in the lower, dextran-rich phase. Ethanol was removed from the top phase by distillation. The alcohol-free top phase was then mixed with the bottom phase, a new sugar solution was added to the flask, and the process was repeated. Ten cycles were completed. The total yield of ethanol per liter was 1556 g from 3407 g glucose, thus, giving a weight conversion of 46%, or 90% of the theoretical yield.

Andersson et al. [199] performed repeated batch processes in PEG/dextran systems using *Bacillus subtilis* for production of α-amylase. The bacterial cells partitioned totally to the bottom phase, and 73% to 82% of the α-amylase partitioned to the top phase. The enzyme concentration reached 0.85 to 1.35 U ml$^{-1}$ in the two-phase mode compared with 0.58 U ml$^{-1}$ in a reference process.

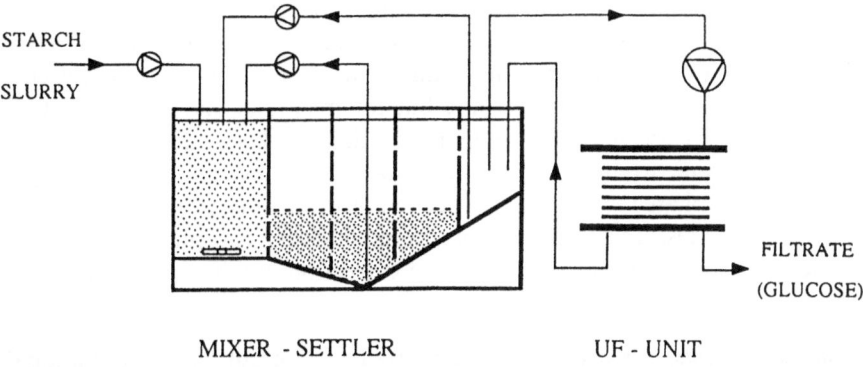

STARCH
SLURRY

FILTRATE

(GLUCOSE)

MIXER - SETTLER                    UF - UNIT

**Fig. 18.** Experimental design for continuous hydrolysis of starch in an aqueous two-phase system. From Larsson et al. [191]

Kaul and Matiasson [195] investigated the conversion of hydrocortisone to prednisolone using *Arthrobacter simplex* in PEG/dextran and PEG/hydroxypropyl starch systems. The steroids were found to prefer the upper phase and the bacterial cells preferred the lower phase. The use of aqueous two-phase systems made it possible to operate the conversion reaction at higher substrate concentrations than in pure buffered solutions. The prednisolone in the top phase could be removed by batch or chromatographic adsorption on Amberlite XAD-4 and elution with methanol.

For the most part, bioconversions have been performed in PEG/dextran or polymer/polymer aqueous two phase systems. One of the first successful attempts at using a PEG/potassium phosphate system was for the production of acrylamide from acrylonitrile using *Brevibacterium* CH1 [183]. The bacterial cells partitioned in the bottom phase ($K = 0.01$), the acrylonitrile ($K = 2.1$) preferred the top phase, and the acrylamide ($K = 1.4$) was more evenly distributed. These partition coefficients were found to reduce inhibition of the active bacterial enzyme by both the substrate and product. The conversion was run five times by repeated use of the bacterial cells in the bottom phase.

# 6 Conclusions

This paper has addressed the theoretical, experimental and practical aspects of aqueous two-phase technology. Aqueous two-phase systems provide a technically and economically effective means for producing biotechnology products. The key advantages of the systems include the ease and reliability with which they can be scaled up, the selectivity they can achieve, and the fact that the polymers comprising the systems stabilize biomolecular structure and activity. A combination of theoretical and experimental work has led to the development of rules and thermodynamic relations for describing and predicting phase separation and biomolecule partition, and thus, will facilitate the widespread use of aqueous two-phase technology. The versatility of the systems has been demonstrated by the fact that they can be used not only for biomolecule purification (down-stream processing), but for microbial and enzymatic bioconversions. Large-scale biomolecule purification has been effectively demonstrated using one to three equilibrium stages with continuous centrifuges performing the phase separation. Extractive bioconversions provide an exciting area for further research and development in which the two-phase systems are used for the simultaneous production and extraction of a biomolecule.

*Acknowledgements*: This work was supported by a grant from the National Science Foundation, a National Institutes of Health Trainee Fellowship, and an Air Products Research Fellowship.

# 7 References

1. Beijerinck MW (1986) Zentralbl Bakteriol Parasitenkd Infektionskr Abt 2 2:627, 698
2. Beijerinck MW (1910) Kolloid-Z. 7:16
3. Dobry A, Boyer-Kawenoki F (1947) J Polym Sci 2:90
4. Craig LC, Craig D (1956) In: Weissberger (ed) Techniques of organic chemistry, vol 4, part 1. Interscience, New York
5. Albertsson P-Å (1985) History of aqueous polymer two-phase systems. In: Walter H, Brooks DE, Fisher D (eds) Partitioning in aqueous two-phase systems. Theory, methods, uses, and applications to biotechnology. Academic, Orlando, p 1
6. Albertsson P-Å (1986) Partition of cell particles and macromolecules, 3rd edn, Wiley, New York
7. Albertsson P-Å (1971) Partition of cell particles and macromolecules, 2nd edn, Wiley, New York
8. Albertsson P-Å (1958) Biochim Biophys Acta 27:378
9. Tjerneld F, Berner S, Cajarville A, Johansson G (1986) Enz Microb Technol 8:417
10. Mattiasson B, Ling TGI (1986) J Chromatogr 376:235
11. Nguyen A-L, Grothe S, Luong JHT (1988) Appl Microbiol Biotechnol 27:341
12. Dobry A (1938) J Chim Phys 35:87
13. Dobry A (1939) J Chim Phys 36:102
14. Tjerneld F (1989) New Polymers for Aqueous Two-Phase Systems. In: Fisher D, Sutherland IA (eds) Separations Using Aqueous Phase Systems. Applications in Cell Biology and Biotechnology. Plenum, New York, p 429
15. Hsu JT (1990) US Patent 4,980,065
16. Stewart R, Topp P (1990) A Polyethylene Glycol-Sodium Chloride Multiphase System for Extraction of Acid Hydrolysates. In: Sikdar SK, Bier M, Topp P (eds) Fronteirs in Bio-processing II, ACS Conference Series II. ACS, Washington, DC
17. Walter H, Brooks DE, Fisher D (eds) (1985) Partitioning in Aqueous Two-Phase Systems. Theory, Methods, Uses, and Applications to Biotechnology. Academic, Orlando
18. Fisher D, Sutherland IA (eds) (1989) Separations Using Aqueous Phase Systems. Applications in Cell Biology and Biotechnology. Plenum, New York
19. Shanbhag VP, Axelsson CG (1975) Eur J Biochem 60:17
20. Walter H, Krob EJ, Brooks DE (1976) Biochemistry 15:2959
21. Zaslavsky BYu, Miheeva LM, Rogozhin SV (1979) Biochim Biophys Acta 588:89
22. Zaslavsky BYu, Miheeva LM, Mestechkina NM, Rogozhin SV (1982) J Chromatogr 240:21
23. Zaslavsky BYu, Miheeva LM, Mestechkina NM, Rogozhin SV (1982) J Chromatogr 253:139
24. Zaslavsky BYu, Mestechkina NM, Rogozhin SV (1983) J Chromatogr 260:329
25. Hustedt H, Kroner KH, Stach W, Kula M-R (1978) Biotechnol Bioeng 20:1989
26. Kroner KH, Hustedt H, Kula M-R (1982) Biotechnol Bioeng 24:1015
27. Kroner KH, Hustedt H, Kula M-R (1984) Process Biochem 19:170
28. Tjerneld F, Johansson G, Joelsson M (1987) Biotechnol Bioeng 30:809
29. Kroner KH, Schütte H, Stach W, Kula M-R (1982) J Chem Tech Biotechnol 32:130
30. Diamond AD, Hsu JT (1989) Biotechnol Bioeng 34:1000
31. Diamond AD, Hsu JT (1989) Biotechnol Techn 3:119
32. Forciniti D, Hall CK (1991) Fluid Phase Equilib 61:243
33. Lei X, Diamond AD, Hsu JT (1990) J Chem Eng Data 35:420
34. Brooks DE, Sharp KA, Fisher D (1985) Theoretical Aspects of Partitioning. In: Walter H, Brooks DE, Fisher D (eds) Partitioning in Aqueous Two-Phase Systems. Theory, Methods, Uses, and Applications to Biotechnology. Academic, Orlando, p 11
35. Albertsson P-Å, Cajarville A, Brooks DE, Tjerneld F (1987) Biochim Biophys Acta 926:87
36. Diamond AD, Hsu JT (1990) AIChE J 36:1017
37. Flory PJ (1941) J Chem Phys 9:660
38. Huggins ML (1941) J Chem Phys 9:440
39. Flory PJ (1953) Principles of Polymer Chemistry. Cornell University Press, Ithaca, New York
40. Zaslavsky BYu, Bagirov TO, Borovskaya AA, Gulaeva ND, Miheeva LH, Mahmudov AU, Rodnikova MN (1989) Polymer 30:2104
41. Gustafsson Å, Wennerström H (1986) Polymer 27:1768
42. Baskir JN, Hatton TA, Suter UW (1987) Macromolecules 20:1300

43. King RS, Blanch HW, Prausnitz JM (1988) AIChE J 34: 1585
44. Kang CH, Sandler SI (1988) Macromolecules 21: 3088
45. Cabezas H, Evans JD, Szlag DC (1990) ACS Symp Ser 419: 38
46. Forciniti D, Hall CK (1990) ACS Symp Ser 419: 53
47. Diamond AD (1990) Fundamental Studies of Biomolecule Partitioning in Aqueous Polymer Two-Phase Systems. Lehigh University, Pennsylvania
48. Diamond AD, Hsu JT (1992) AIChE Symp Ser (in press)
49. Johansson G (1970) Biochim Biophys Acta 221: 387
50. Bamberger S, Seaman GVF, Brown JA, Brooks DE (1984) J Colloid Interface Sci 99: 187
51. Zaslavsky BYu, Bagirov TO, Borovskaya AA, Gasanova GZ, Gulaeva ND, Levin VYu, Masimov AA, Mahmudov AU, Mestechkina NM, Miheeva LM, Osipov NM, Rogozhin SV (1986) Colloid Polymer Sci 264: 1066
52. Zaslavsky BYu, Mahmudov AU, Bagirov TO, Borovskaya AA, Gasanova GV, Gulaeva ND, Levin VYu, Mestechkina NM, Miheeva LM, Rodnikova MN (1987) Colloid Polymer Sci 265: 548
53. Zaslavsky BYu, Miheeva LM, Aleschko-Ozhevskii YuP, Mahmudov AU, Bagirov TO, Garaev ES (1988) J Chromatogr 439: 267
54. Abrams DS, Prausnitz JM (1975) AIChE J 21: 116
55. Scheutjens JMHM, Fleer GJ (1979) J Phys Chem 83: 1619
56. Scheutjens JMHM, Fleer GJ (1980) J Phys Chem 84: 178
57. Edmond E, Ogston AG (1968) Biochem J 109: 569
58. Hill TL (1959) J Chem Phys 30: 93
59. Baskir JN, Hatton TA, Suter UW (1989) Biotechnol Bioeng 34: 541
60. Abbot NL, Blankschtein D, Hatton TA (1990) Bioseparation 1: 191
61. Diamond AD, Hsu JT (1990) J Chromatogr 513: 137
62. Johansson G, Andersson M (1984) J Chromatogr 303: 39
63. Johansson G (1985) Partitioning of Proteins. In: Walter H, Brooks DE, Fisher D (eds) Partitioning in Aqueous Two-Phase Systems. Theory, Methods, Uses, and Applications to Biotechnology. Academic, Orlando, p 161
64. Johansson G, Kopperschläger G, Albertsson P-Å (1983) Eur J Biochem 131: 589
65. Johansson G, Joelsson M, Olde B (1990) Biochim Biophys Acta 1029: 295
66. Albertsson P-Å, Nyns EJ (1961) Ark Kemi 17: 197
67. Albertsson P-Å, Sasakawa S, Walter H (1970) Nature 228: 1329
68. Johansson G (1971) Affects of Different Ions on the Partition of Proteins in an Aqueous Dextran-Poly(ethylene) Glycol Two-Phase System. Proceedings of the Int'l Solvent Extraction Conference, 1971, The Hague, 2: 928
69. Walter H, Sasakawa S, Albertsson P-Å (1972) Biochemistry 11: 3880
70. Sasakawa S, Walter H (1972) Biochemistry 11: 2760
71. Sasakawa S, Walter H (1974) Biochemistry 13: 29
72. Gelsema WJ, Deligny CL (1982) Sep Sci Technol 17: 375
73. Johansson G (1985) J Chromatogr 322: 425
74. Davis JT, Rideal EK (1961) Interfacial Phenomena. Academic, London
75. Reitherman R, Flanagan SD, Barondes SH (1973) Biochim Biophys Acta 297: 193
76. Brooks DE, Sharp KA, Bamberger S, Tamblyn CH, Seaman GVF, Walter H (1984) J Colloid Interface Sci 102: 1
77. Johansson G (1970) Biochim Biophys Acta 222: 381
78. Johansson G, Hartman A, Albertsson P-Å (1973) Eur J Biochem 33: 379
79. Hughes P, Lowe CR (1988) Enz Microb Technol 10: 115
80. Cheng L, Joelsson M, Johansson G (1990) J Chromatogr 523: 119
81. Shanbhag VP, Johansson G (1974) Biochem Biophys Res Communications 61: 1141
82. Shanbhag VP, Axelsson C-G (1975) Eur J Biochem 60: 17
83. Johansson G (1976) Biochem Biophys Acta 451: 517
84. Axelsson C-G (1978) Biochim Biophys Acta 533: 34
85. Shanbhag VP, Johansson G (1979) Eur J Biochem 93: 363
86. Johansson G, Shanbhag VP (1984) J Chromatogr 284: 63
87. Johansson G, Tjerneld F (1989) J Biotechnol 11: 135
88. Menge U, Morr M, Mayr U, Kula M-R (1983) J Appl Biochem 5: 75
89. Takerkart G, Segard E, Minsigny M (1974) FEBS Lett 42: 218
90. Andrews BA, Head DM, Dunthorne P, Asenjo JA (1990) Biotechnol Techn 4: 49
91. Flanagan SD, Barondes SH (1975) J Biol Chem 250: 1484

92. Axelsson CG, Shanbhag VP (1976) Eur J Biochem 71:419
93. Hubert P, Dellacherie E, Neel J, Baulieu E-E (1976) FEBS Lett 65:169
94. Kula M-R, Johansson G, Buchman AF (1979) Biochem Soc Transact 7:1
95. Cordes A, Kula M-R (1986) J Chromatogr 376:375
96. Erlanson-Albertsson C (1980) FEBS Lett 117:295
97. Pinaev G, Tartakovsky A, Shanbhag VP, Johansson G, Backman L (1982) Mol Cell Biochem 48:65
98. Kopperschläger G, Johansson G (1982) Anal Biochem 124:117
99. Johansson G, Kopperschläger G, Albertsson P-Å (1983) Eur J Biochem 131:589
100. Johansson G, Andersson M, Åkerlund H-E (1984) J Chromatogr 298:483
101. Kopperschläger G, Lorenz G, Usbeck E (1983) J Chromatogr 259:97
102. Johansson G, Joelsson M (1985) Biotechnol Bioeng 27:621
103. Johansson G, Joelsson M, Åkerlund H-E (1985) J Biotechnol 2:225
104. Johansson G, Joelsson M (1986) Appl Biochem Biotechnol 13:15
105. Tjerneld F, Johansson G, Joelsson M (1987) Biotechnol Bioeng 30:809
106. Johansson G, Joelsson M (1987) J Chromatogr 393:195
107. Johansson G, Joelsson M (1987) J Chromatogr 411:161
108. Johansson G, Joelsson M (1987) J Chromatogr 393:195
109. Birkenmeier G, Usbeck E, Kopperschläger G (1984) Anal Biochem 136:265
110. Birkenmeier G, Tschechoniem B, Kopperschläger G (1984) FEBS Lett 174:162
111. Kroner HK, Cordes A, Schelper A, Morr M, Bückmann AF, Kula M-R (1982) Affinity Partition Studies with Glucose-6-Phosphate Dehydrogenase in Aqueous Two-Phase Systems in Response to Triazine Dyes. In: Gribnau TCJ, Visser J, Nivard RJE (eds) Affinity Chromatography and Related Techniques. Elsevier, Amsterdam
112. Johannson G, Joelsson M (1984) J Chromatogr 291:175
113. Johansson G, Joelsson M (1985) Enz Microb Technol 7:629
114. Persson LO, Olde B (1988) J Chromatogr 457:183
115. Shiemann J, Kopperschläger G (1984) Plant Sci Lett 36:205
116. Chen J-P, Carlson A (1987) Biochnol Techn 1:263
117. Plunket SD, Arnold FH (1990) Biotechnol Techn 4:45
118. Lee C-K, Sandler SI (1990) Biotechnol Bioeng 35:408
119. Wuenschell GE, Naranjs E, Arnold FH (1990) Bioprocess Eng 5:199
120. Birkenmeier G, Vijayalaksmi MA, Stigbrand T, Kopperschläger G (1991) J Chromatogr 539:267
121. Blanco MTM, Perez JAC, Olde B, Johansson G (1986) J Chromatogr 358:147
122. Johansson G (1986) J Chromatogr 368:309
123. Johansson G (1989) Affinity Partition of Enzymes. In: Fisher D, Sutherlans IA (eds) Separations Using Aqueous Phase Systems. Applications in Cell Biology and Biotechnology. Plenum, New York, p 7
124. Cordes A, Flassdorf J, Kula M-R (1987) Biotechnol Bioeng 30:514
125. Suh S-S, Arnold F (1990) Biotechnol Bioeng 35:682
126. Kroner KH, Hustedt H, Granada S, Kula M-R (1978) Biotechnol Bioeng 20:1967
127. Hustedt H, Kroner KH, Kula M-R (1985) Applications of Phase Partitioning in Biotechnology. In: Walter H, Brooks DE, Fisher D (eds) Partitioning in Aqueous Two-Phase Systems. Theory, Methods, Uses, and Applications to Biotechnology. Academic, Orlando, p 529
128. Yang J, Shen Z, Wang J (1987) Shengwu Gongcheng Zuebao 3:273
129. Hustedt H, Kroner KH, Schütte H, Kula M-R (1983) Extractive Purification of Intracellular Enzymes. In: Enzyme Technol Rotenburg Ferment Symp, 3rd, 1982, p 135
130. Albertsson P-Å, Andersson B (1981) J Chromatogr 215:131
131. Smeds AL, Enfors SO (1985) Enz Microb Technol 7:601
132. Kroner KH, Schütte H, Stach W, Kula M-R (1982) J Chem Technol Biotechnol 32:130
133. Veide A, Smeds A-L, Enfors SO (1983) Biotechnol Bioeng 25:1789
134. Schwenzer B, Kopperschläger G (1989) Biochim Biophys Acta 48:33
135. Hustedt H, Kroner KH, Menge U, Kula M-R (1978) Aqueous Two-Phase Systems for Large-Scale Enzyme Isolation Processes. In: Prepr - Eur Congr Biotechnol, 1st, 1978, Part 1, p 48
136. Delgado C, Tejedor MC, Luque J (1990) J Chromatogr 498:159
137. Hummel W, Schütte H, Kula M-R (1985) Appl Microbiol Biotechnol 21:7
138. Hummel W, Schütte H, Kula M-R (1984) Ann NY Acad Sci 434:195
139. Hummel W, Schütte H, Kula M-R (1983) Eur J Appl Microbiol Biotechnol 18:75
140. Schütte H, Kroner KH, Hummel W, Kula M-R (1983) Ann NY Acad Sci 413:270

141. Albertsson P-Å (1973) Biochemistry 12: 2525
142. Veide A, Strandberg L, Enfors SO (1987) Enz Microb Technol 9: 730
143. Chen J-P (1989) J Food Sci 54: 1369
144. Szlag DC, Guiliano KA (1988) Biotechnol Techn 2: 277
145. Craig LC (1960) Partition. In: Alexander P, Black RJ (eds) A Laboratory Manual of Analytical Methods of Protein Chemistry. Pergamon Press, Oxford, vol 1, p 122
146. Treffry TE, Sharpe PT, Walter H, Brooks DE, (1985) Thin Layer Countercurrent Distribution and Apparatus. In: Walter H, Brooks DE, Fisher D (eds) Partitioning in Aqueous Two-Phase Systems. Theory, Methods, Uses, and Applications to Biotechnology. Academic, Orlando, p 131
147. Albertsson P-Å (1965) Anal Biochem 11: 121
148. Albertsson P-Å (1970) Science Tools 17: 56
149. Åkerlund H-E (1984) J Biophys Biochem Meth 9: 133
150. Blomquist G, Albertsson P-Å (1972) Biochim Biophys Acta 73: 125
151. Treybal RE (1951) In: Liquid Extraction. McGraw-Hill, New York
152. Hustedt H, Kroner KH, Menge U, Kula M-R (1980) Enz Eng 5: 45
153. Kula M-R, Kroner KH, Hustedt H, Schütte H (1981) Ann NY Acad Sci 369: 341
154. Martin AIP, Synge RLM (1941) 35: 1358
155. Morris CJOR (1963) Protides Biol Fluids, Proc Colloq 10: 325
156. Müller W (1986) Kontakte (Darmstadt) 3: 3
157. Müller W, Schuetz H-J, Guerrier-Takada C, Cole PE, Potts R (1979) Nucl Acids Res 7: 2483
158. Müller W, Kutemeier G (1982) Eur J Biochem 128: 231
159. Müller W (1985) Partitioning of Nucleic Acids. In: Walter H, Brooks DE, Fisher D (eds) Partitioning in Aqueous Two-Phase Systems. Theory, Methods, Uses, and Applications to Biotechnology. Academic, Orlando, p 227
160. Müller W (1986) Eur J Biochem 155: 213
161. Müller W (1989) Liquid-Liquid Partition Chromatography of Biopolymers in Aqueous Two-Phase Systems. In: Fisher D, Sutherlans IA (eds) Separations Using Aqueous Phase Systems. Applications in Cell Biology and Biotechnology. Plenum, New York, p 381
162. Heubner A, Juchem M, Müller W, Pollow K (1989) Application of Liquid-Liquid Partition Chromatography (LLPC) in the Preparation of Steroid Binding Proteins. In: Fisher D, Sutherlans IA (eds) Separations Using Aqueous Phase Systems. Applications in Cell Biology and Biotechnology. Plenum, New York, p 393
163. Matsumoto U, Shibusawa Y (1980) J Chromatogr 187: 351
164. Matsumoto U, Shibusawa Y (1981) J Chromatogr 206: 17
165. Matsumoto U, Shibusawa Y, Tanaka Y (1983) J Chromatogr 268: 375
166. Matsumoto U, Ban M, Shibusawa Y (1984) J Chromatogr 285: 69
167. Matsumoto U, Shibusawa Y (1986) J Chromatogr 356: 27
168. Shibusawa Y, Matsumoto U, Takatori M (1987) J Chromatogr 398: 153
169. Ito Y., Weinstein MA, Aoki I, Harada R, Kimura E, Numogaki K (1966) Nature 212: 985
170. Sutherland IA, Ito Y (1978) HRC CC J High Resolut Chromatogr Chromatogr Commun 3: 171
171. Heywood-Waddington D, Sutherland IA, Morris WB, Peters TJ (1984) Biochem J 217: 751
172. Flanagan SD, Johansson G, Yost B, Ito Y, Sutherland IA (1984) J Liq Chromatogr 7: 385
173. Sandlin JL, Ito Y (1988) J Liq Chromatogr 11: 15
174. Ito Y, Bowman RL (1973) Science 182: 391
175. Ito Y (1987) Sep Sci Tech 22: 1971
176. Sutherland IA, Ito Y (1980) Anal Biochem 108: 367
177. Ito Y, Bramblett GT, Bhatnagar R, Humberman M, Leive LL, Culliname LM, Groves W (1983) Sep Sci Technol 18: 33
178. Leive LL, Culliname LM, Ito Y, Bramblett GT (1984) J Liq Chromatogr 7: 403
179. Hsu JT, Chou FE (1991) Protein Separation Using Aqueous Two-Phase Systems in the Eccentric Multi-Layer Coil Planet Centrifuge. In: 7th International Conference on Partitioning in Aqueous Two-Phase Systems. Advances in Separation in Biochemistry, Cell Biology and Biotechnology. New Orleans, Louisiana, number 83
180. Lee YW, Cook CE, Chen DC, Ray P, Shibusawa Y, Ito Y (1991) Application of High Speed Countercurrent Chromatography/Aqueous twp-Phase Solvent System for the Purification of Recombinant Proteins. In: 7th International Conference on Partitioning in Aqueous Two-Phase Systems. Advances in Separation in Biochemistry, Cell Biology and Biotechnology. New Orleans, Louisiana, number 81

181. Shibusawa Y, Ito Y, Lee YW (1991) Countercurrent Chromatography of Proteins with Polymer Phase Systems Using Cross-Axis Synchronous Coil Planet Centrifuge. In: 7th International Conference on Partitioning in Aqueous Two-Phase Systems. Advances in Separation in Biochemistry, Cell Biology and Biotechnology. New Orleans, Louisiana, number 82
182. Andersson E, Hahn-Hägerdal B (1990) Enz Microb Technol 12: 242
183. Lee YH, Chang HN (1989) Biotechnol Lett 1: 23
184. Wennersten R, Tjerneld F, Larsson M, Mattiasson B (1985) Proc Int Solvent Extraction Conf ISEC Denver, p 506
185. Kuhlmann W, Halwachs W, Schügerl K (1980) Chem Ing Tech 52: 607 Synopse 821
186. Tjerneld F, Persson I, Albertsson P-Å, Hahn-Hägerdal B (1985) Biotechnol Bioeng 27: 1036
187. Tjerneld F, Persson I, Albertsson P-Å, Hahn-Hägerdal B (1985) Biotechnol Bioeng 27: 1044
188. Tjerneld F, Persson I, Lee JM (1991) Biotechnol Bioeng 37: 876
189. Nguyen A-L, Grothe S, Luong JHT (1988) Appl Microbiol Biotechnol 27: 341
190. Larsson M, Mattiasson B (1988) Biotechnol Bioeng 31: 979
191. Larsson M, Arasaratnam V, Mattiasson B (1989) Biotechnol Bioeng 33: 758
192. Hayashida K, Kunimoto K, Shiraishi F, Kawakami K, Arai Y (1990) J Ferment Bioeng 69: 240
193. Yamazaki Y, Suzuki H (1979) Rep Ferment Res Inst Japan 52: 33
194. Andersson E, Mattiasson B, Hahn-Hägerdal B (1984) Enz Microb Technol 6: 301
195. Kaul R, Mattiasson B (1986) Appl Microbiol Biotechnol 24: 259
196. Persson I, Tjerneld F, Hahn-Hägerdal B (1989) Biotechnol Techn 3: 265
197. Andersson E, Hahn-Hägerdal B (1988) Appl Microbiol Biotechnol 29: 329
198. Yong Hee L, Ho Nam C (1990) J Ferment Bioeng 69: 89
199. Andersson E, Johansson A-C, Hahn-Hägerdal B (1985) Enz Microb Technol 7: 333
200. Kantelinen A, Poutanen K, Linko M, Markkanen P (1987) Ann NY Acad Sci
201. Kantelinen A, Poutanen K, Linko M, Markkanen P (1987) In: Laskin AI, Mosbach K, Thomas D, Wingard LB Jr (eds) Enzyme Engineering 8. Plenum, New York
202. Yang LW, Hustedt H, Kula M-R (1988) Biotechnol Appl Biochem 10: 173
203. Mattiasson B, Souminen M, Andersson E, Häggström L, Albertsson P-Å, Hahn-Hägerdal B (1982) In: Chibata I, Fukui S, Wingard LB Jr (eds) Enzyme Engineering 6. Plenum, New York
204. Puziss M, Hedén C-G (1965) Biotechnol Bioeng 7: 355
205. Flygare S (1988) Ph.D. Thesis, Dept of Pure and Appl Biochemistry, Lund University, Sweden
206. Kühn I (1980) Biotechnol Bioeng 22: 2393
207. Hahn-Hägerdal B, Mattiasson B, Albertsson P-Å (1981) Biotechnol Lett 3: 53
208. Larsson M, Mattiasson B (1984) Ann NY Acad Sci 434: 144
209. Larrson M, Mattiasson B (1984) In: Laskin AI, Tsao GT, Wingard LB Jr (eds) Enzyme Engineering 7. Plenum Publishing Corp, New York
210. Persson I, Tjerneld F, Hahn-Hägerdal B (1984) Enz Microb Technol 6: 415

# Novel Separations Based on Affinity Interactions

J.H.T. Luong and A-L. Nguyen
Biotechnology Research Institute National Research Council Canada,
Montreal, Quebec, Canada H4P 2R2

Novel purification processes have been developed, based on the affinity interaction between complementary molecules, to circumvent the difficulties associated with affinity chromatography. Depending upon the procedure used for isolating the ligand-ligate complex, the process can be termed affinity ultrafiltration, affinity partitioning or affinity precipitation. This review describes the developments, potentials, and applications of such purification procedures. Emphasis is also placed on the type and choice of affinity ligands as well as some common procedures by which ligands can be covalently immobilized to water-soluble or insoluble matrices. Finally, the problems and challenges encountered by such novel purification procedures are presented and discussed.

Advances in Biochemical Engineering
Biotechnology, Vol. 47
Managing Editor:
© Springer-Verlag Berlin Heidelberg 1992

# 1 Introduction

Recent breakthroughs in recombinant DNA (rDNA), genetic engineering, and cell-fusion techniques make the production of various proteins/enzymes and specialty chemicals available. The feasibility of using rDNA technology to stimulate the extracellular secretion of protein also offers further opportunities for cost reduction and simplification of the purification processes. The manufacture of such products is of interest in the food, agricultural, pharmaceutical and chemical industries. So far, one of the major bottlenecks toward commercialization has been the slow development of techniques for isolating and purifying fragile bioproducts. Not only are these valuable compounds vulnerable to enzymatic degradation and physical damages such as pH, temperature, shear and solvent used in purification, but they are often present in low concentrations in complex mixtures containing other similar materials as well. In many cases, the desired products are often unstable, especially in the presence of impurities. Such characteristics inevitably pose several technical problems for the design and operating conditions of the purification technique [1].

Purification from complex biological mixtures has traditionally been performed by combining techniques which resolve substances according to differences in their overall physico-chemical properties. Extensive purification thus translates into a high yield loss and therefore a high production cost. In many rDNA bioprocesses, the purification of the protein product can contribute up to 90% of the overall processing cost [2]. New, robust purification procedures which possess both high resolution and recovery are still needed for very dilute and delicate products. This is the most critical area demanding innovative process development and intensive research.

For the last decade, impressive progress has been achieved in purification technology by exploiting the natural affinity displayed between biochemicals and their complementary ligands. Affinity chromatography was first introduced by Campbell et al. [3] in 1951 and Lerman [4] in 1953 as a method for separation of anti-hapten antibodies and tyrosinase, respectively. This technique was then refined by Cuatrecasas et al. [5] and by Porath et al. [6].

Affinity chromatography can achieve yields as high as 90% and a purification factor of over 6000 [7, 8]. This technique which revolutionized protein/enzyme purification in the 1970s, is the second most common purification method used at the laboratory scale, after ion-exchange chromatography. In this procedure, an affinity matrix is prepared by chemically binding a complementary ligand (sometimes called the affinant), often via a suitable spacer molecule or leash, to the surface of insoluble bead materials. Only the biomolecule (called ligate or binder) which recognizes the immobile ligand will be retained on the matrix while other molecules pass by unretarded. Subsequently, a solution containing an appropriate eluent can be passed through the column to release and recover the bound product.

Affinity chromatography offers high selectivity, especially when a mono-clonal antibody is used as the affinity ligand (immunoadsorbent chromato-graphy). This technique, however, does have some major drawbacks. The system can only be operated batchwise and it uses a packed column that can become plugged and fouled. Consequently, pretreatment of the feed to remove solid contaminants is a prerequisite. Because of the low throughput, affinity chro-matography is a very low-productivity system: its processing rate can be as low as $10\,\mathrm{kg\,h^{-1}}$ compared with that of precipitation procedures ($> 1000\,\mathrm{kg\,h^{-1}}$; [1]). Owing partly to very expensive ligands and the unsuitability of many matrices available in commercial quantities, affinity chromatography has been almost exclusively used for the preparation of the very costly products such as urokinase [9], interferon, and antithrombin III. In addition, the scale-up of this process is very complicated and the results have not been well documented. On the commercial scale, affinity chromatography makes up only 3% of the current bioseparation procedures but it is expected to increase to 8% within a decade. Currently, the market for the separation of biologically active materials is worth about $\$770 \times 10^6$ and is expected to be valued at $\$2.2 \times 10^9$ in 1993 [10]. Some major biotechnology companies are now selling and/or utilizing products for commercial scale affinity separations.

Affinity interactions can, however, be used rather more subtly to circumvent the difficulties encountered by conventional chromatography. Depending upon the procedure used for isolating the ligand-ligate complex, the process can be termed affinity ultrafiltration, affinity partition or affinity precipitation. Recent efforts have focused on the development of affinity ultrafiltration by several research groups and the possibility of combining affinity interactions with precipitation or aqueous two-phase partitioning has also been explored. This chapter reviews the principles, applications and potentials of such novel puri-fication techniques. Development trends in affinity interactions for the pre-paration of enzymes/proteins are also presented and discussed.

## 2 General Considerations in Affinity Interactions

### 2.1 Principles of Affinity Interactions

The affinity interaction displayed between a biological macromolecule (M); namely a ligate and its complementary ligand (L), sometimes called the affinant, or effector, is a well-known phenomenon:

$$M + L \underset{k_{-1}}{\overset{k_1}{\rightleftharpoons}} ML \tag{1}$$

The dissociation constant, $K_p$ [the ratio between the dissociation ($k_{-1}$) and association or binding ($k_1$), rate constants] is defined as:

$$K_p = \frac{k_{-1}}{k_1} = \frac{[M][L]}{[ML]} \tag{2}$$

where [L], [M] and [ML] are the concentration of ligand (L), ligate (M), and complex (ML), respectively. A low $K_p$ thus represents an extremely stable complex and vice versa. Typical values are $K_p = 10^{-15}$ M for the biotin-avidin interaction, $K_p = 10^{-6} - 10^{-12}$ M for antigen-antibody, and $10^{-4} - 10^{-6}$ M for lectin-carbohydrate and enzyme-substrate interactions [11].

In conventional affinity chromatography, the maximum immobilized ligand concentration attainable ($L_o$) is about 5–10 mM for low molecular weight ligands [12], and for significant retardation of the ligate, $L_o$ must be significantly higher than $K_p$. In practice, a ratio between $L_o/K_p$ of 10 is considered satisfactory, thus if the ligand concentration is $10^{-2}$ M, $K_p$ must be smaller than $10^{-3}$ M for reasonably effective adsorption.

## 2.2 Type and Choice of Ligand

Ligands can be classified into two main groups according to their specificity: highly specific and group specific.

### 2.2.1 Highly Specific Ligands

A wide range of small molecules (MW < 1000) has been used as highly specific ligands. Such ligands include enzyme substrates, substrate analogs, enzyme inhibitors, allosteric effectors, nucleic acids, t-RNA, plant hormones, steroids, chromophores, antibiotics etc. For such ligands, the dissociation constant ($K_p$) varies considerably from $> 10^{-3}$ M for substrate analogs to $10^{-15}$ M for the avidin-biotin complex.

Highly specific macroligands consist of antibody/antigen, proteins, and polysomes. The antigen-antibody binding is probably the most specific interaction known in biology with the dissociation constant $K_p$ varying from $10^{-6}$ to $10^{-2}$ M. Besides antigen/antibody, a few examples of proteins used as specific ligands are protein A and protein G for isolation of immunoglobulins and cells, lectins for isolation of polysaccharides, oligosaccharides, glycoproteins and membrane proteins, α-acid glycoprotein for isolation of cationic and anionic drugs, gelatin for isolation of fibronectin, insulin for isolation of insulin receptor etc. A detailed compilation of proteins as specific ligands can be found elsewhere [13].

## 2.2.2 Group Specific Ligands

Several ligands are known to interact more or less specifically with groups of macromolecules. Enzyme cofactors and analogs such as NAD/NADH, AMP, and other nucleotides are called group-specific since more than one ligate will bind to the same ligand, e.g. $NAD^+$ by dehydrogenases. Heparin, a complex polysaccharide consisting of glucose and glucosamine interacts with many enzymes such as lipoprotein lipase, DNA and RNA polymerases, restriction endonuclease, collagenase, iduronidase, hyaluronidase, heparin N-sulfatase, and trehalose phosphate synthase. This complex polysaccharide also interacts with several proteins and blood proteins including thrombin, antithrombin III, factor IX, factor XI, factor VII, and high and low density proteins.

Another useful set of group specific ligands are reactive synthetic dyes mainly based on the triazine structure. Some of the best known reactive dyes are Cibacron Blue 3G-A, Procion Red H-3B, Procion Yellow H-A, Procion Scarlet MX-G etc. About 90 reactive dyes are commercially available in large quantities and detailed information on such reactive dyes can be found elsewhere [14]. In general, they are very stable, inexpensive, and mimic some of the naturally occurring nucleotide-containing cofactors such as NAD/NADH, AMP, ATP etc. Consequently, the majority of dehydrogenases, kinases, t-RNA synthases, and other enzymes interacting with nucleotides are expected to bind these reactive compounds. Proteins such as albumin, lipoproteins, neoantigens, ferritin, ricin, interferon, blood clotting factors, histidine-rich glycoproteins, and receptors (transferrin, estradiol, vitamin D) are also known to interact with the reactive dyes. The reactive dyes are frequently referred to as pseudo-affinity ligands.

Boric acid is another important group specific ligand. To date, immobilized boronates have been commercially available and used for purifying carbohydrates, nucleosides, nucleotides, nucleic acids, proteins, and small molecules such as citric acid, tyrosine, pyridoxal, lactic acid, and norepinephrine.

## 2.2.3 Choice of Ligand

In order to exploit affinity interactions for a given purification, the following conditions must be fulfilled:

1. The dissociation constant ($K_p$) should be less than $10^{-3}$ M. This rule of thumb, established experimentally by several investigators, was confirmed theoretically by Graves and Wu [15]. In addition, this constant must not be smaller than $10^{-8}$ M, otherwise dissociation of the complex becomes very difficult, particularly if either the ligand or the ligate has only limited stability, e.g. antigen-antibody interaction.
2. The ligand should be bifunctional, i.e. it offers a site for specific affinity binding and a separate site for immobilization. This is clearly more prob-

lematical with small ligands since they possess few functional groups in comparison to macroligands.

3. The ligand must be stable during immobilization and under operating conditions.
4. The ligand should contain no hydrophobic or charged groups since such groups can cause nonspecific adsorption of protein.
5. The ligand must be specific for a target compound. In general, specificity in binding requires that at least one of the participating species possesses high molecular weight, for instance, antigen-antibody interaction. In addition, very specific ligands may be unstable, costly to prepare, and toxic.
6. Feasibility for selectively desorbing the bound substance from the ligand without denaturing the biological activity of the purified product and the ligand, i.e., the complex must be reversible. This requirement may be problematical for antibody/antigen purification because the complex formed is very stable. As a matter of fact, the harsh conditions, e.g. extremes of pH (2–3) or 6 M urea applied for removing the binding substance from the immunoadsorbent, frequently resulted in denaturing of the ligand and/or the desired product.
7. The ligand should be inexpensive, non-toxic and available in large quantities. This may be the main disadvantage for ligands such as antibodies, proteins, and cofactors, since they are costly and denature easily and quickly.

In practice, the choice of a suitable ligand is dictated by the cost and quality of purified product, the nature of crude extracts, the method used for purification, and the intended use of the purified product. The antigen/antibody interaction is highly specific but this type of ligand is very expensive, unstable, and must be produced for a specific application. In principle, it is possible to produce any kind of antibody, however, by no means is this a trivial task as to whether the antibody is polyclonal or monoclonal. Reactive dyes are inexpensive, stable, available in industrial quantities and not biodegradable. However, the resolution of such pseudo-ligands is much lower than that of immunosorption and the problems related to toxicity of reactive dyes remain unresolved. This issue is of importance if the purified product is used for injection.

# 3 Affinity Ultrafiltration

## 3.1 Principles of Affinity Ultrafiltration

Ultrafiltration is a separation process based on the ability of a membrane to pass some solution components and retard others. The method is efficient, economical, and simple since it involves no phase change, no chemical addition, and very low energy requirements. The greatest disadvantage of ultrafiltration, however, is its low resolution; a 10-fold difference in molecular weight is about

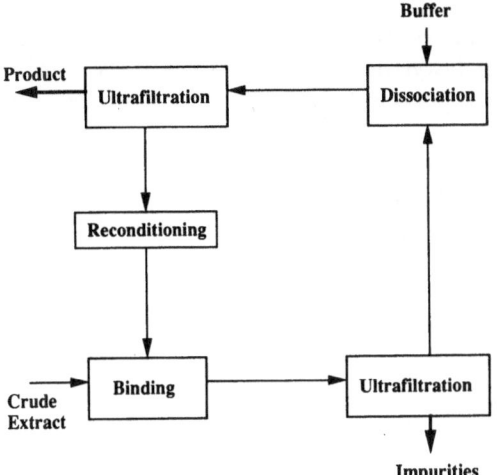

**Fig. 1.** Schematic diagram of affinity ultrafiltration

the best that can be expected [1]. High resolution can be achieved by exploiting affinity interaction in conjunction with ultrafiltration and a method called affinity ultrafiltration has been developed. Its basic principle is that the substance to be purified (ligate or binder) when present in a crude mixture, will pass through the membrane. The binder, however, will be retained by the membrane when it binds to a very high molecular weight ligand, while other unbound components of the mixture will pass through. The isolated ligand-binder complex is then treated with an appropriate eluent to desorb the binder from the ligand. The ligand is recycled and reconditioned if necessary (Fig. 1).

If a homogenate contains no particulate matter, a crude solution of the binder is mixed with the ligand and affinity binding takes place. Alternatively, in a more general approach which is applicable for continuous purification of material from a bioreactor, for instance, the crude solution and the macroligand are delivered separately to either side of a suitable molecular weight cut-off membrane. The binding occurs after the substance to be purified passes through the membrane.

## 3.2 Preparation of the Macroligand

Macroligands are not always available for a given purification and in such cases, the preparation of the macroligand is a prerequisite. For application in affinity ultrafiltration, the ligand can be attached to a water-insoluble or soluble matrix to form a macroligand. Like affinity chromatography, the matrix used in affinity ultrafiltration must have the following properties:

1. It should contain no intrinsic charged groups in order to minimize non-specific ionic adsorption of enzymes/proteins.

2. It should contain no hydrophobic groups such as nonpolar side arms (R groups) since proteins very frequently adsorb nonspecifically and irreversibly to hydrophobic surfaces such as polystyrene beads. This could be the reason why hydrophilic matrices such as polysaccharides are almost exclusively utilized for affinity chromatography.
3. It should possess a sufficient number of active groups which could be activated/modified to form covalent binding with the ligand.
4. It must be mechanically and chemically stable during modification and to the operation conditions of affinity ultrafiltration.
5. It must be resistant to microorganisms, non-toxic, inexpensive and available in large quantities.

### 3.2.1 Attachment of the Ligand to Water-Insoluble Matrices

In affinity chromatography, hydrophilic polymers which have been frequently used as matrix materials are polyacrylamide, polysaccharides such as cellulose, agarose, dextran and porous glass beads.

Agarose, a polygalactan obtained from seaweed, is one of the most used matrices because of its high porosity and chemical versatility. However, agarose disintegrates in organic solvents and at high temperature, therefore, this type of material cannot be used in preparation processes which require sterilization.

Cellulose is chemically and physically stable and can be used for preparations of affinity adsorbents. In conventional affinity chromatography, cellulose is not widely used since it is neither microporous nor homogeneous. Recently, bead-form cellulose was prepared which appeared to be highly porous, physically rigid and chemically versatile [16].

Sephadex, one of the commercially available dextran matrices, is a cross-linked product of dextran and epichlorohydrin. As compared with agarose, it is also chemically versatile and highly porous but this gel is mechanically inferior.

Polyacrylamides possess a marked hydrophilic character and are very resistant to biodegradation. The low porosity of polyacrylamides, however, prevents this gel from being a good matrix in affinity chromatography. In addition, polyacrylamide generates ionic interactions with enzyme/proteins, particularly at alkaline pH since it contains carboxyl groups produced from the hydrolysis of its amide groups.

Derivatized porous glass beads such as aminopropyl, arylamine and isothiocyanato glass are frequently used because of their rigidity and high porosity. This type of matrix, however, has less binding capacity and silican in the glass bead generates a nonspecific adsorption effect (anionic at neutral pH because of its silica constituents) and becomes less stable in alkaline medium.

Similar to enzyme immobilization, various techniques are available for the attachment of the ligand to water-insoluble matrices. The chemistry involved for the coupling of the ligand to polyacrylamides and glass beads is somewhat similar to that for polysaccharides although the activation procedure is different.

The activation procedure for agarose can be adapted easily for cellulose. To date, polysaccharides are by far the most popular matrices used for affinity purifications since they can easily be modified/activated.

The bisoxirane (bisepoxide) method, e.g. 1:4-butanediol diglycidoxyether is commonly used for the activation of polysaccharide since bisoxirane reacts with the hydroxyl groups of the matrix to form an activated gel containing free epoxy group available for coupling reactions. In this case, the ligand must possess a nucleophilic moiety such as an amino, thiol, or hydroxyl group. The main drawback of this method is the extreme difficulty of removing unreacted oxirane groups [17]. In addition, the coupling step may require several days although 24 h are generally sufficient. Coupling efficiency can be enhanced significantly by increasing pH to the range of 8–12 or by an increase in temperature to 40 °C. Obviously, such a condition may not be acceptable for some protein ligands. The advantage of this method is its relative non-toxicity and the formation of stable bonds between the matrix and the ligand.

The activation of polysaccharide gels, including agarose, by cyanogen bromide is another popular method especially for the attachment of antibody ligands [17]. At alkaline pH 11–12 and at low temperature, cyanogen bromide reacts rapidly with the hydroxyl groups of such gels to form a cyclic major product which is believed to be imidocarbonate. The coupling reaction between imidocarbonate and the amine of the ligand appears to generate several products including isourea, $N$-substituted imidocarbonate and $N$-substituted carbamate. Besides the high toxicity of cyanogen bromide, the isourea linkage exists in the protonated cationic form and may promote the ionic adsorption effect, which is nonspecific and usually considered as an interference. In addition, isourea bonds formed on coupling amine-containing ligands are not very stable, so ligand leakage is another drawback of this coupling procedure.

The periodate method is another common procedure which offers a rapid and simple attachment of the ligand. In this method, sodium periodate is used to oxidize the *cis*-vicinal hydroxyl groups of carbohydrate to aldehyde which in turn can react directly with a primary amine of the ligand to form a Schiff's base [17]. To improve binding, the aldehyde can be reacted with a bifunctional dihydrazide to form a hydrazone/hydrazido derivative. This derivative is then reduced with sodium borohydride or sodium cyanohydridoborate to form a hydrazine derivative which can be used for coupling with the carboxyl or the carbonyl group of the ligand. Alternatively, the hydrazido group can react with nitrous acid to form an azido derivative which can bond covalently with the primary amine of the ligand. This method has proved very useful for coupling nucleic acids and nucleotides including $NAD^+/NADP$. The disadvantage of this method is the requirement of cyanohydridoborate, a toxic compound. The periodate reaction is difficult to control and the reaction with amines is rather slow.

After covalent binding of the ligand, it is important to block the remaining unreacted active groups. Otherwise, the substrate to be purified may form covalent linkage with the matrix. Blocking agents such as glycine and 2-amino-

2-(hydroxymethyl)-1,3-propanediol are widely used. The carbodiimide method can be used to block unreacted amino or carboxyl groups with low molecular weight substances, e.g. 2-amino-ethanol.

### 3.2.2 The Use of a Spacer-Arm (Leash)

The concept of using a spacer-arm was introduced by Baker in 1967 [18] to improve the ligand-binder interaction and N-alkyl moieties are commonly used to space the ligand from the matrix, particularly for low molecular weight ligands. In many cases, the introduction of the spacer-arm is useful and mandatory since it allows an increased flexibility and mobility of the ligand so that efficient protein/enzyme adsorption can take place. Alternatively, the leash structure can be attached to the ligand and the complex then coupled to the matrix. Although this strategy minimizes the number of unreacted spacer-arms, it requires more preparation steps and skillful labor.

In order to minimize localized steric hindrance and nonspecific hydrophobic adsorption, it is advantageous to use spacer-arms of a more hydrophilic nature by incorporating polar groups such as carbinol, amino or secondary amino groups to the leash structure [19]. In addition, the indiscriminate use of long spacer-arms in the preparation of the affinity matrix should be avoided. In general, systems with high $K_p$ values require longer spacer-arms than those with lower $K_p$ values. The incorporation of the spacer-arm to the matrix may also limit the amount of affinity ligand due to the hydrophobic leash structure.

### 3.2.3 Preparation of Water-Soluble Macromolecular Ligands

When particle-bound ligands are used, there is a limit to the amount of binder that can be bound and the introduction of a leash structure may be required to improve the ligand-ligate interaction. Such drawbacks are well recognized from experiments in affinity chromatography.

For applications using affinity ultrafiltration, the ligand could be attached to a water-soluble polymer directly to form a macroligand. The homogeneous character of the system is expected to facilitate the ligand-binder interaction to the extent that the binding process could occur almost instantaneously without the requirement of the spacer-arm.

A few dextran derivatives such as octadecylamino dextran, octadecylamide dextran, and N-amino dextran should be quite useful for attaching a variety of affinity ligands [20]. Luong et al. [21] synthesized an affinity polymer by copolymerizing acrylamide and N-acryloyl-m-aminobenzamidine in the absence of oxygen. This procedure may be adapted for the preparation of other affinity polymers. Guzman et al. [22] described the procedure for covalent attachment of various enzymes to water-soluble nonionic surfactants such as ethoxylated fatty alcohols or acids. Unlike the ionic surfactants, these materials exhibited a negligible effect on the activity of protein even at a concentration as high as $10^{-3}$ M. The attachment of the ligand is reversible and 100 times higher than covalent binding of ligand to agarose gel.

## 3.3 Applications of Affinity Ultrafiltration

The affinity ultrafiltration technique was applied to purify concanavalin A from a crude extract of jack beans (*Canavalia ensiformis*) by using heat-killed cells of *Saccharomyces cerevisiae* as the affinity adsorbent [23]. Using 0.8 M glucose as the eluent, a highly purified product was obtained with an overall yield of 70%. The interaction between concanavalin A (MW = 102,000) and heat-killed cells (mean diameter of 5 μm) occurred in a mixing chamber, washing and dissociation were performed in membrane units with molecular mass cut-off of 300–1000 kDa. Consequently, concanavalin A in free form passed through the membrane. However, the concanavalin A-heat killed cells complex, was large enough and could not permeate.

Affinity ultrafiltration was also applied to purify alcohol dehydrogenase (ADH) from a crude extract of *S. cerevisiae* [24]. Cibacron Blue, a dye that binds to NAD-and ATP-requiring enzymes, was immobilized to starch granules to form a macroligand. After centrifugation to remove all particulate matters, a yeast homogenate was mixed with the macroligand. In this process, a hollow fiber unit with a molecular mass cut-off of 500,000 was used to isolate the macroligand-binder complex which was then dissociated with a potassium phosphate solution (0.55 M). Neither product purity nor yield were reported in this work.

Choe et al. [25] attempted to attach a soybean trypsin inhibitor to dextran and used this affinity polymer for purification of trypsin from a mixture of chymotrypsin and trypsin. The overall yield and the purity of the recovered trypsin was 55% and 81%, respectively. Similarly, Adamski-Medda et al. [26] purified trypsin from a chymotrypsin-trypsin mixture using dextran-*p*-amino-benzamidine. Although the recovery yield of this process was higher (76%), the purity of recovered trypsin was considerably lower (65%). Apparently, dextran bound chymotrypsin nonspecifically and adversely affected the purity of the purified trypsin.

Affinity ultrafiltration was extended to purify β-galactosidase from cell homogenates of *Escherichia coli* [27] using commercially available *p*-aminobenzyl-1-thio-β-D-galactopyranoside agarose beads. The process consists of two identical stages of adsorbing and desorbing where the affinity beads were kept in the external compartments of two containers equipped with concentric cylindrical filtration devices (20 μm membrane filter). The feed stream containing the crude β-galactosidase and the wash buffer were mixed and fed to the adsorption unit as one stream. The resulting affinity-adsorbent binder complex was then transferred to the desorbing stage where dissociation took place at pH 10. The containers were agitated on a rotary shaker in order to keep the affinity beads in suspension. The suspension of affinity adsorbent was recirculated between the external compartments of the two containers by a multichannel peristaltic pump. The recovery yield per ml gel of this continuous purification process was reported to be 70% with a corresponding productivity of $25.2 \text{ U ml}^{-1} \text{h}^{-1}$.

Trypsin was purified from chymotrypsin using a water-soluble polymer having an acrylamide backbone with $m$-aminobenzamidine as pendants [28]. In batch, the procedure gave 90% yield and 98% purity. Operating continuously, and using benzamidine as the eluent because of its specificity, an overall yield of 45% of highly purified trypsin from porcine pancreatic extract was accomplished [29]. The trypsin-binding capacity of the affinity polymer approached the theoretical value of 120:1 (weight of bound trypsin:weight of aminobenzamidine present), which is considerably higher than that of commercially available gel matrices (insoluble). This affinity polymer was also used for purification of urokinase, a plasminogen-activating proteolytic enzyme, from human urine [30]. The process yield was determined to be 49% and the recovered urokinase exhibited a specific activity close to that of the highest commercial grade. Noteworthy is the fact that acrylamide and aminobenzamidine were used because they were able to withstand the high shear force, temperature, pH and other harsh operating conditions.

## 3.4 Critique on Affinity Ultrafiltration

Similar to conventional chromatography, the affinity ultrafiltration process has the advantage of selective separation of one substance that has an affinity for a ligand from other substances which may have the same molecular weight but different affinities to the ligand. The high resolution and recovery possible with affinity ultrafiltration, along with its capability of processing unclarified and viscous liquids, have given a large impetus to developing its application further. The technique can be applied to the direct purification of biochemicals from the spent medium.

In developing a particular process, the choice of ligand is important, since properties other than its specificity will affect the efficiency and operating life of the system. Attachment of ligand to suitable macromolecules can be accomplished by many well-established procedures as described earlier. The invulnerability of chemical (pseudo) ligands to the shear force, temperature, pH etc., means they are easy to handle and possess a long operating life. However, suitably specific chemical ligands are not always available in which case, biochemical affinity, e.g. antibody-antigen interaction, can be exploited. However, handling difficulty in view of dissociation, reusability of the ligand etc. may arise. Presently, both types of ligands are subjects of intensive investigations. In the past, the technological limit to the use of membranes in separation processes was the lack of suitable membranes in terms of molecular mass cut-offs, flux, and operating life. Today, however, membranes are available with high resistance to acids, bases, alcohols, and temperature, which allow very effective cleaning and sanitization. Possibilities also exist for fabricating membranes with high molecular mass cut-offs between $10^5$ to $10^6$, a region of great interest for protein separations and such membranes are commercially available from Amicon, Abcor, Millipore, Dorr-Oliver, Ultra-Pore, Nucleopore etc. [1]. To a large extent, the applicability of affinity ultrafiltration for purification of enzymes/ proteins will depend on the successful development of inexpensive, reliable, and robust membranes.

The nonspecific adsorption of the affinity macroligand and/or proteins to the membrane could result in a significant reduction in the permeation rate and may form a dynamic membrane with its own rejection characteristics. This problem could be circumvented by adding electrostatic charge to the membrane and/or the macroligand. This is an important and fascinating area in affinity ultrafiltration, a subject to which greater research effort should be devoted.

Non-specific adsorption of proteins to the affinity macroligand is another matter of concern since this phenomenon is rather common and unpredictable. In general, protein adsorption is entropically driven and dependent upon the physical and chemical properties of proteins, the adsorbent surface and the type of solvent used. Since proteins are polyelectrolytic in nature, protein adsorption due to ionic interaction will occur if the affinity ligand contains ionic groups which may come from the matrix, ligand, spacer arm or the coupling agent, e.g. cyanogen bromide. Maximum adsorption is attained at the isoelectric point and some proteins can adsorb strongly even to surfaces with the same charge as the proteins [31]. An understanding of protein adsorption is, therefore, important in the preparation of the affinity ligand and/or the membrane used for separation of biomolecules of interest.

Affinity ultrafiltration has been used mainly on a laboratory scale but the technique is being investigated intensively to overcome inherent drawbacks. Its principle can be easily applied in scaled-up preparation and this promising technique is likely to be used in the not-too-distant-future, at least on a small production scale.

# 4  Affinity Partitioning

## 4.1  Formation of Aqueous Two-Phase Systems

A solution of two polymers in an organic solvent often settles into two phases where each polymer is predominantly found in one phase [32]. This demixing phenomenon does not generate any practical interest. Similarly, a long period of neglect followed the first observation of the phase formation by polymers in aqueous solutions. These aqueous two-phase systems only began to attract a great deal of attention after their inherent biocompatibility and their potential for large scale applications were recognized [33].

The spontaneous phase formation was first reported for the systems of soluble starch and gelatin, presently the systems most often studied and used, contain polyethylene glycol (PEG) and dextran [33]. Quantitative data concerning the phase composition can be presented in a three-component phase diagram similar to that used for liquid-liquid extraction. However, a triangular diagram for an aqueous two-phase system always displays the phase composition envelope very close to the corner corresponding to high water percentages, in marked contrast to a typical three-component liquid system. The truncated diagram presenting only the corner for high water content, affords a better perception of changes in the phase compositions but can appear misleading to readers accustomed to the full triangular diagrams. Therefore, data on aqueous two-phase systems are customarily presented using rectangular coordinates with the understanding that the water content can be calculated from the polymer concentrations (Fig. 2).

The popularity of the PEG/dextran systems is due to the relative suitability of the physico-chemical properties of the polymers, their biodegradability and their non-toxicity. Since dextran is rather expensive, other polymers including hydroxypropyl starch [34], pullulan [35], and polyvinylalcohol [36] have been studied as potential substitutes. Generally, the phase diagram depends strongly on the polymer molecular weight and the molecular weight distribution. Moreover, adding salt to the system also alters the phase diagram remarkably. In fact, with a sufficient concentration of salt, a solution of PEG alone also settles to a PEG-predominant phase and a salt phase.

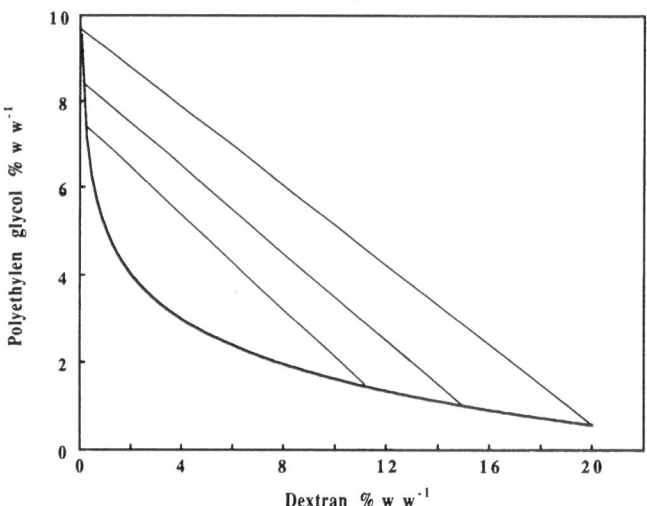

**Fig. 2.** A typical phase diagram of a dextran-polyethylene glycol system

When low molecular weight compounds, such as inorganic salts, sugars, amino acids, and nucleotides are introduced to an aqueous two-phase system they generally distribute evenly between the two phases. However, poly-electrolytes such as DEAE-dextran, exhibit a preference for one phase. Higher molecular weight materials, including proteins, nucleic acids, cell organelles, membrane vesicles, and whole cells display remarkable partition in many cases. In PEG-containing systems, the PEG-predominant phase always has a lower specific gravity. It is therefore often referred to as the upper or top phase. The dextran, or salt-predominant phase is designated as the lower or bottom phase. For a partitioned species, the ratio of concentration in the upper phase over that in the lower phase is known as the partition coefficient. Table 1 presents the partition coefficients of some representative substances, in some typical systems.

Solution thermodynamic theories have been used to explain the phase formation and the partition of substances between the phases [37, 38]. The economics of the process, including the feasibility of polymer reutilization and the environmental impact have also been studied [39]. Moreover, the engineer-ing aspects of unit operation, such as multistage design, counter-current ar-rangement, scale-up etc. have also been extensively investigated [40]. Above all, the partition coefficient is the prime parameter in the economics of the partition-ing process. This coefficient is the result of the interaction between the phase components and the partitioned species. The interaction includes the electro-chemical, hydrophobic, and conformational effects, none of which can be predicted. The partition coefficients therefore must be experimentally deter-mined for the target in each system. However, if a phase-forming polymer

**Table 1.** The partition coefficient of selected substances in liquid-liquid two phase systems [33]

| Partitioned substances | Partition coefficient | Dext (w/w %) | PEG (w/w %) | Added salt (M) |
|---|---|---|---|---|
| LiI | 1.11 | 7 | 7 | — |
| $K_2SO_4$ | 0.84 | 7 | 7 | — |
| Citric acid | 1.44 | 7 | 7 | — |
| DEAE dextran | 13.6 | 5 | 4 | 0.01 (NaI) |
| CM dextran | 0.12 | 5 | 4 | 0.01 (NaI) |
| Phycoerythrin | 7 | 7 | 4 | 0.1 (Na-citrate) |
| DNA | 30 | 5 | 4 | 0.01 (LiCl) |
| DNA | 0.015 | 5 | 4 | 0.01 (KCl) |
| Chloroplast | 9 | 6 | 6 | — |
| *E. coli* | 9 | 6 | 6 | — |
| β-galactosidase | 62* | | | |

\* partitioned in PEG-salt systems
 DEAE dextran:diethylaminoethyl dextran
 CM dextran:carboxymethyl dextran

possesses moieties (ligands) that have specific affinity for a partitioned substance, the latter will prefer the phase containing more of the ligands. Aqueous
two-phase processes in which a phase-forming polymer has been specifically
modified to contain such a ligand are referred to as affinity partitioning.

## 4.2 Synthesis of Ligand-Bearing Polymers

All the ligands useful in affinity chromatography and affinity ultrafiltration are
also used for affinity partitioning. Therefore, inhibitors [41] and cofactors [42]
have been bound to polymers for enzyme partitioning. In addition, the metal
chelating groups [43] and the protein binding dyes [44, 45] have also been
studied. In principle, it is possible to derivatize PEG and dextran as well as the
dextran substitutes. In fact, many of the dextran derivatives that were cited for
affinity ultrafiltration applications can also be used with PEG to form aqueous
two-phase systems. A comprehensive review [20] and an updated listing are
available elsewhere [46].

The same publications also reviewed methods for the derivatization of PEG
which has attracted even more attention. Since PEG is quite inert chemically,
the terminal hydroxyl groups must be substituted with some active groups such
as halides, sulfonate esters, or epoxides before reacting with the ligand. Since
PEG derivatives are also used to synthesize PEG-bound proteins for other
application, novel PEG derivatization procedures continue to be reported and
will further extend possibilities for affinity partitioning.

## 4.3 Applications of Affinity Partitioning

The first report of affinity partitioning, an example of enzyme-inhibitor interaction, concerned the purification of trypsin using PEG-bound p-aminobenzamidine [47]. The affinity interaction induced 92% of the trypsin to be
partitioned in the PEG phase. In another early study, a case of hydrophobic
affinity, palmitic acid was bound to PEG, causing 90% of albumin in human
plasma to be partitioned in the upper phase while other proteins were restricted
to the lower dextran phase [48]. Some other studies used estradiol and lecithin
to purify oxosteroid isomerase and colipase, respectively [49, 50]. The stability
and the low prices of the reactive dyes make them very attractive for affinity
partitioning. They have been used to purify phosphofructokinase, glucose-6-
phosphate dehydrogenase, and other enzymes [51]. The synthesis of cofactor-
bearing PEG is laborious and often requires the introduction of spacer-arm
between the polymer and the ligand. For example, $N^6$-(2-aminoethyl)-NADH-
PEG has been prepared and was reported to effect the affinity partitioning of the
hydrogenases for lactate, alcohol, formate, and formaldehyde [52].

All the foregoing applications utilized PEG-bound ligands while dextran-
bound ligands were rarely studied. In one report, ATP-adipoyl dihydrazo-

dextran was prepared and used to partition phosphoglycerate kinase from spinach chloroplasts [42]. After an ion-exchange treatment the enzyme preparation was 80% pure.

## 4.4 Critique on Affinity Partitioning

As mentioned earlier, the aqueous two-phase systems provide a gentle environment for biochemicals and affinity partitioning inherits this advantage. As in the case of affinity ultrafiltration, the ligand-target interaction also occurs in the liquid state and can be expected to proceed expeditiously. The greatest concern about affinity partitioning lies in the cost of the ligand-bearing polymer. Therefore, much effort has been focused on the recycling of the PEG-bound ligands. Recycling of dextran-bound ligands is somewhat more difficult due to their high viscosity, and to date, no report has appeared.

A recent study reported the use of Cibacron Blue bound to Sepharose particles in conjunction with a PEG-dextran two-phase system [53]. The purification of pyruvate kinase was conducted, followed by enzyme recovery with the addition of salt. This process permitted reutilization of the ligand-bearing entity, however, only a two-fold purification was achieved.

In another direction, a water soluble polymer was synthesized from $N$-isopropylacrylamide and glycidyl acrylate [54]. The glycidyl residues permitted the copolymer to react with the amino groups of $p$-aminobenzamidine. This affinity polymer was found to form aqueous two-phase systems with dextran or pullulan and was able to bind trypsin, thus effecting a trypsin partition. Owing to the $N$-isopropyl acrylamide units, the affinity polymer precipitated from the isolated upper phase, upon addition of ammonium sulfate at low pH. This condition also induced the affinity polymer to release trypsin to the unbound state. The affinity polymer could be reused but there was a significant loss during each cycle of reutilization.

Affinity partitioning has a great potential for the large-scale purification of biochemicals. Although the procedures for preparing affinity polymers are readily available, the processes for recovery and reutilization must be fully developed before this technique becomes economically viable.

# 5  Affinity Precipitation

## 5.1 Principles of Affinity Precipitation

Another promising affinity separation technique involving precipitation of the ligand-product complex is defined as affinity precipitation. In one approach, a ligand and a precipitation group are co-attached to a water-soluble polymer and the resulting macroligand can be precipitated by temperature, pH and/or salinity variation. The second approach makes use of bi- or polyvalents that bind to specific sites of enzymes/proteins and promote the formation of insoluble complexes.

## 5.2 Applications of Affinity Precipitation

Affinity precipitation was applied for purifying trypsin from bovine pancreas by Schneider et al. [55]. A water-soluble polyacrylamide was developed which bore a ligand group (p-aminobenzamidine) and a precipitation group (benzoic acid). The affinity polymer was added directly to a crude extract under conditions favoring the binding of the desired protein and was then precipitated at pH 3. After removing the supernatant, the protein of interest was eluted from the affinity polymer which was then recycled. The recovery yield was 90% and the recovered trypsin was essentially pure and active. In this process, the amount of polymer added to the medium was very low (0.1–0.5%). Thus, non-specific adsorption of proteins on the polymer was also low and the affinity polymer was recycled for repeated use without losing its binding capacity. The average polymer loss per cycle was about 1%. In general, this approach may cause some problems particularly with proteins or enzymes which are vulnerable at low pH.

Precipitation by temperature and/or salinity variation mitigates this short-coming. Poly-N-isopropylacrylamide (NIPAM) is a well studied polymer whose solubility is dependent upon temperature and salinity. This property has been exploited to produce recoverable and reusable enzyme conjugates and to develop a novel immunoassay procedure [56]. In the latter development, the polymer was conjugated with a monoclonal antibody. After contacting with the antigen, the triplex antigen-antibody-polymer was precipitated by raising the temperature to 31 °C.

Exploitation of thermal phase separation for purification was realized by copolymerizing N-isopropyl acrylamide (NIPAM) with N-acryloxysuccinimide (NASI) or glycidyl methacrylate (GMA). The amino groups of ligands were able to react with the residues of NASI or GMA and the polymers were precipitated by temperature and/or salinity variation, since they contained the NIPAM residues [57]. Such reactive polymers were reacted with p-aminobenzamidine, and used for purification of trypsin from a trypsin-chymotripsin mixture with an overall yield of 82%. The NIPAM-GMA copolymer was also reacted with immunogammaglobulin (IgG) and the IgG bearing polymer was able to bind protein A to form a precipitable complex [57].

The first investigation of affinity precipitation utilized a bifunctional nucleotide derivative, $N_2,N_2$-adipohydrazido-bis-($N^6$-carbonylmethyl-NAD) (Bis-NAD) to precipitate the tetrameric enzyme lactate dehydrogenase (LDH) [58]. Since LDH is a tetrameric enzyme, Bis-NAD interacted with two LDH molecules to form large aggregates and the solution eventually became insoluble when these aggregates were sufficiently large. The overall recovery yield was reported to be 85%. Following this approach, Larsson et al. [59] and Pearson et al. [60] synthesized affinity precipitation reagents using Bis-NAD and bistriazine dyes to crosslink lactate dehydrogenase. Both the dyes and NAD are known to bind LDH at specific sites because of the enzyme's oligomeric structure. Affinity precipitation was also utilized to isolate erythrocytes with $N,N'$-bis-3-(dihydroxyborylbenzene)-adipamide. This ligand probably displays the lectin-like property of agglutinating red blood cells [61]. A heterofunctional ligand was also prepared by attaching soybean trypsin inhibitor to chitosan and used for purifying trypsin [62]. The chitosan part of the affinity ligand was used for precipitation since it readily precipitated at pH 8–9.

Bis-chelates of Cu(II) were used for selective precipitation of horse cytochrome c, sperm whale myoglobin and human hemoglobin [63]. The rationale for the choice of these proteins is the variety of histidines on the protein's surface (26 for human hemoglobin, 6 for sperm whale myoglobin, and one partially accessible for cytochrome c) and the binding between Cu(II) and the metal-coordinating residues on the protein's surface. The precipitation appeared to depend upon the Cu/histidine ratio and pH. The nature and activity of the proteins precipitated were not reported.

## 5.3 Critique on Affinity Precipitation

Affinity precipitation offers the most economical use of immobilized ligands for rapid and specific isolation of enzymes/proteins on a large scale. This technique could be extended to purify enzymes/proteins providing suitable bifunctional ligands are available. The concept of using affinity precipitation is very attractive because it is possible to reduce the volume of the sample handled in later purification steps, thus reducing the overall cost of purification. Noteworthy is the fact that besides dead-end complex formation, other ternary complexes were able to be developed, e.g. complexes with coenzyme and inhibitor, or with coenzyme-substrate adducts. The formation of ternary complexes is particularly useful when the interaction between the bifunctional ligand and the enzyme is not sufficiently strong. To date, the development of this technique has been slow compared to that of affinity ultrafiltration and affinity partitioning. This is due in part to the limited availability of precipitable affinity polymers, the use of very expensive bis-ligands and to the undetermined effect of interfering compounds present in crude extracts. However, it seems possible that this technique will find considerable application in the large-scale purification of enzymes/proteins and other specialty chemicals.

# 6  Development Trends and Conclusions

The unique specificity and reversibility of biological interactions have opened a new horizon for the development of purification technologies. In their qualities of scale, resolution, and recovery yield, the novel techniques based on affinity interactions have the potential to replace existing process technologies such as affinity chromatography and high performance liquid chromatography.

Traditionally, affinity interactions have been reserved for the final steps of protein/enzyme purification. These techniques, however, undoubtedly can be used in the initial phase of purification schemes because of their high resolution as well as high recovery yield. The following important aspects of affinity interactions must be fully addressed before such purification techniques become commercially viable.

1. The cost of ligands still presents a significant problem, especially with protein ligands such as antibodies. Even though the costs of antibody production are decreasing continuously, the use of an antibody as a specific ligand is still mainly used for the preparation of high value products.
2. Chemical modification of an antigen/antibody to reduce the dissociation constant $(K_p)$ thus facilitating elution is the logical way to overcome excessive affinity. Surprisingly, this research area has remained untouched. The recombinant DNA and hybridoma methods could provide the ligands of weak affinity interaction $(K_p = 10^{-4} - 10^{-2} \text{ M})$; [64].
3. The development of low cost pseudo-ligands such as reactive dyes should be explored intensively. The possible toxicity of reactive dyes is not yet solved adequately and this issue must be addressed properly. In addition, an understanding of the interaction between a reactive dye with a specific protein is another crucial step which requires intensive research and development. Nevertheless, such dyes and other group specific ligands including heparin, lectins, proteins A and G are expected to be widely used for the purification of lower cost proteins.
4. Another hurdle yet to be overcome is the method of ligand attachment. Although a plethora of attachment chemistries have been reported, very few have been investigated thoroughly. In addition, many of these procedures are covered by patents making commercial ventures based on affinity interactions more difficult.
5. The leakage of ligand from the matrix on prolonged use must be thoroughly investigated, especially for the purification of pharmaceutical products which are administered by injection.
6. Sterility of the purification process is of importance if purified products are aimed for use in pharmaceuticals, especially those for injection. Heat sterilization is not suitable for several matrices and protein ligands and chemical sterilization must be applied. Some triazine dyes can be heat sterilized and this is another distinct advantage of these pseudo-ligands.
7. Attention must be focused on the nature of the purified protein, i.e. monitoring its activity and stability during the purification process.
8. Last but not least, work is needed to generate predictive models of the affinity process in order to facilitate scale-up.

In addition to the preparation of enzymes/proteins and specialty chemicals, novel procedures based on affinity interactions are also expected to be used

**Table 2.** Some possible applications of novel separation techniques based on affinity interactions

| Target compound | Ligand | Application | Ref |
|---|---|---|---|
| Pyrogen | Polymixin B | Removal of pyrogens in solutions | [65] |
| Virus | Hepatitis B surface antigen | Removal of Hepatitis from plasma | [66] |
| Undesired antibody | DNA | Removal of anti-DNA antibodies | [67] |
| Bilirubin | Serum albumin | Removal of bilirubin in blood | [68] |
| DL-Aldosterone | Corticosteroid | Resolve optical isomers | [69] |
| Lipids | Decylamine-agarose | Removal of lipoproteins from plasma | [70] |
| Lymphocytes T & B | Antigen/antibody | Fractionation of lymphocytes into T & B cell sub-types | [71] |

widely for the removal of trace contaminants such as pyrogens, viruses, undesired antibodies, and toxins (Table 2). Because of their scale-up potential, high resolution, and high recovery yield, the above purification techniques would appear to have good prospects for large scale industrial use.

The key word of successful commercial biotechnology is purification and this importance is noted by Genentech, the world leading biotechnology company: "Laboratory creation of a new product is the first step. It only begins to count when the product is purified and packaged".

# 7 References

1. Luong JHT, Nguyen AL, Male KB (1987) Trends Biotechnol 5:281
2. Dwyer JL (1984) Bio/Technology 2:957
3. Campbell DH, Lenscher EL, Lerman LS (1951) Proc Natl Acad Sci USA 37:575
4. Lerman LS (1953) Nature 172:635
5. Cuatrecasas P, Wilchek M, Anfinsen CB (1968) Proc Natl Acad Sci USA 61:636
6. Porath J, Axen R, Ernback S (1967) Nature 215:1491
7. Cuatrecasas P, Wilchek M (1968) Biochem Biophys Res Commun 33:235
8. Allen RH, Majerus (1972) J Bio Chem 247:7702
9. Herion P, Bohlen A (1983) Biosci Rep 3:373
10. Sii D, Sadana A (1991) J Biotechnol 19:83
11. Lowe CR (1984) J Biotechnol 1:3
12. Harvey MJ, Lowe CR, Craven DB, Dean PDG (1974) Eur J Biochem 41:335
13. Wilcheck M, Trayer IP, Kohn J (1984) Meth Enzymol 104C:3
14. Stead CV (1987) The chemistry of reactive dyes. In: Clonis YD, Atkinson A, Bruton CJ, Lowe CR (eds) Reactive dyes in protein and enzyme technology, Stockton Press, NY, p 13
15. Graves DJ, Wu YT (1974) Meth Enzymol 34:140
16. Chen LF, Tsao GT (1977) Biotechnol Bioeng 19:1463
17. Scouten WH (1981) Affinity chromatography: bioselective adsorption on inert matrices, Wiley-Interscience, NY
18. Baker BR (1967) Design of active-site directed irreversible enzyme inhibitor, Wiley, NY, p 301
19. O'Cara P, Barry S, Griffin T (1974) Meth Enzymol 34:108
20. Harris JM, Yalpani M (1986) Polymer-ligands used in affinity partitioning and their synthesis partitioning in aqueous two-phase systems. In: Walter H, Brooks DE, Fisher D (eds) Academic Press, Orlando

21. Luong JHT, Male KB, Nguyen AL (1988) Biotechnol Bioeng 31:439
22. Guzman R, Torres Jl, Carbonell RG, Kilpatrick PK (1989) Biotechnol Bioeng 33:1267
23. Mattiasson B, Ramstorp M (1984) J Chromatogr 283:322
24. Mattiasson B, Ling TGI (1986) Ultrafiltration affinity purification: a process for large-scale biospecific separations. In: McGregor WC (ed) Membrane separation in biotechnology, Marcel Dekker, NY
25. Choe TP, Masse P, Verdier A (1986) Biotechnol Lett 3:163
26. Adamski-Medda D, Nguyen QT, Dellacherie E (1981) J Membr Sci 9:337
27. Pungor E, Afeyan NB, Gordon NF, Cooney CL (1987) Bio/Technology 5:604
28. Male KB, Luong JHT, Nguyen AL (1987) Enz Microbial Technol 9:374
29. Luong JHT, Male KB, Nguyen AL (1988) Biotechnol Bioeng 31:516
30. Male KB, Nguyen AL, Luong JHT (1990) Biotechnol Bioeng 35:87
31. Norde W, MacRitchie F, Nowicka FG, Lyklema J (1986) J Colloid Interface Sci 112:447
32. Kern RJ (1956) J Poly Sci 21:49
33. Albertsson P-A (1986) Partition of cells particles and macromolecules, Wiley, NY
34. Tjerneldt F, Berner S, Cajarville A, Johansson G (1986) Enz Microbial Technol 8:417
35. Nguyen AL, Grothe S, Luong JHT (1988) Appl Microbiol Biotechnol 27:341
36. Hughes P, Lowe CR (1988) Enz Microbial Technol 10:15
37. Brooks D, Sharp K, Fisher D (1985) Theoretical aspects of partitioning. In: Walter H, Brooks DE, Fisher D (eds) Partitioning in aqueous two-phase systems, Academic Press, Orlando
38. Sanchez IC, Balazs AC (1989) Macromolecules 22:2325
39. Hustedt H (1986) Biotechnol Lett 8:791
40. Hustedt H, Kroner KH, Kula M-R (1985) Applications of phase partitioning in biotechnology. In: Walter H, Brooks DE and Fisher D (eds) Partitioning in aqueous two-phase systems, Academic Pres, Orlando
41. Chen J-P, Carlson A (1987) Biotechnol Techniques 1:263
42. Persson L-O, Oldes B (1988) J Chromatogr 457:183
43. Porath J, Carlsson J, Olsson I, Belfrase G (1975) Nature 258:598
44. Birkenmeier G, Usbeck E, Kopperschläger G (1984) Anal Chem 136:264
45. Cordes A, Flossdorf J, Kula M-R (1986) Biotechnol Bioeng 30:514
46. Kopperschläger G, Birkenmeier G (1990) Bioseparation 1:235
47. Takerkart G, Segard E, Monsigny M (1974) FEBS Lett 42:218
48. Shanbag VP, Johansson G (1984) Biochem Biophys Res Commun 61:1141
49. Hubert P, Dellacherie E, Neel J, Beaulieu EE (1976) FEBS Lett 65:169
50. Erlanson-Albertsson C (1980) FEBS Lett 117:295
51. Johansson G, Andersson M (1984) J Chromatogr 291:175
52. Buckmann AF, Morr M, Kula M-R (1987) Biotechnol Appl Biochem 9:258
53. Ku C-A, Henry JD Jr, Blair JB (1989) Biotechnol Bioeng 33:1081 and 1089
54. Nguyen AL, Luong JHT (1990) Enz Microbial Technol 12:663
55. Schneider M, Guillot C, Lamy B (1981) Ann NY Acad Sci USA 369:257
56. Monji N, Hoffman AS (1987) Appl Biochem Biotechnol 14:107
57. Nguyen AL, Luong JHT (1989) Biotechnol Bioeng 34:1186
58. Larsson PO, Mosbach K (1979) FEBS Lett 98:333
59. Larsson PO, Flygare S, Mosbach K (1984) Meth Enzymol 104:364
60. Pearson JC, Burton SJ, Lowe CR (1986) Anal Biochem 158:382
61. Burnett TJ, Peebles HC, Hageman JH (1980) Biochem Biophys Res Commun 96:157
62. Senstad C, Mattiasson B (1989) Biotechnol Bioeng 33:216
63. Van Dam ME, Wuenschell GS, Arnold FH (1989) Biotechnol Appl Biochem 11:492
64. Ohlson S, Nilsson R, Niss U, Kjelberg BM, Freiburghaus C (1988) J Immunol Meth 114:175
65. Duff GW, Waisman DM, Atkins E (1982) Clin Res 30:565A
66. Arensen PDW, Charm SE, Wong BL (1980) Biotechnol Bioeng 22:2207
67. Habib RE, Coulet PR, Sanhadji K, Gautheron DC, Laville M, Treager J (1984) Biotechnol Bioeng 26:665
68. Plotz PH, Berk PD, Scharshmidt BF, Gordon JK, Vegalla J (1974) J Clin Invest 53:778
69. Varsano-Aharen N, Ulick S (1972) J Biol Chem 247:4939
70. Deutsch DG, Fogleman DJ, Von Kaulla (1973) Biochem Biophys Res Commun 50:758
71. Menderino GL, Gooch GT, Stavitsky AB (1978) Cell Immunol 41:264

# Selective Precipitation

M. Q. Niederauer and C. E. Glatz
Department of Chemical Engineering, Iowa State University, Ames,
Iowa 50011, USA

Precipitation methods which offer specificity towards a target protein can greatly decrease the number of steps required to achieve product purity. Techniques which offer a high degree of specificity include the use of polyelectrolytes, biospecific affinity ligands, metal ion affinity ligands, and protein-binding dyes. This article reviews the current technology involving these selective precipitants and the mechanisms by which binding and precipitation occur. In addition to the choice of precipitant, factors such as the manner of precipitant addition, mixing, and the nature of the crude protein mixture also influence the selectivity. Furthermore, the selectivity of these precipitants can often be enhanced through the use of genetic engineering on the target protein. Techniques allowing for removal and recovery of the precipitate are discussed. Comparisons of the selectivities achieved by "selective precipitation" are made with other downstream purification methods such as liquid chromatography. The advantages and disadvantages of using precipitation versus liquid chromatography are reviewed. Novel applications of selective precipitants are presented.

Advances in Biochemical Engineering
Biotechnology, Vol. 47
Managing Editor:
© Springer-Verlag Berlin Heidelberg 1992

## List of Symbols and Abbreviations

| | |
|---|---|
| A | Empirical constant |
| AS | Hydroxypropyl methylcellulose acetate succinate |
| B | Empirical constant |
| C | Empirical constant |
| Con A | Concanavalin A |
| CMC | Carboxymethyl cellulose |
| DEAE | Diethylaminoethyl |
| DMPE-B | Dimyristoylphosphatidylethanolamine-biotin |
| EGTA | Ethylene glycol bis (β-aminoethyl ether)$N,N'$-tetraacetic acid |
| I | Ionic strength |
| K | Equilibrium dissociation constant for binding |
| LDH | Lactose dehydrogenase |
| GMA | Glycidyl methacrylate |
| IgG | Immunogammaglobulin |
| LDH | Lactose dehydrogenase |
| NAD | Nicotinamide adenine dinucleotide |
| NADH | Nicotinamide adenine dinucleotide, reduced form |
| NASI | $N$-acryloxysuccinimide |
| NIPAM | $N$-isopropyl acrylamide |
| PAA | Poly(acrylic acid) |
| PAB | $p$-aminobenzamide |
| PEI | Polyethyleneimine |
| PEG | Polyethylene glycol |
| Poly GAB | PAB covalently attached to the NIPAM-GMA copolymer |
| Poly SAB | PAB covalently attached to the NIPAM-NASI copolymer |
| STI | Soy trypsin inhibitor |
| WBA | Tungstoboric acid |
| WGA | Wheat germ agglutinin |
| WPA | Tungstophosphoric acid |
| Z | Net charge |
| α | Separation factor |
| β | Purification factor |

*Subscripts*

| | |
|---|---|
| b | Binding site |
| c | Concentration |
| p | Protein |
| m | Michaelis-Menton |

# 1 Introduction

Precipitation is one of the primary methods used to achieve concentration during product recovery. In common practice, precipitation is the separation technique most often used during the early stages of downstream processing to achieve partial purification of the product as well as a reduction in volume [1, 2]. Since the product is most often found in the precipitate, both enrichment and concentration are accomplished in one step. Concentration reduces costs because smaller volumes go on to further processing. The precipitating agents which have greater selectivity tend to be more expensive, but may be recyclable. Precipitation processes can be scaled up, are readily suitable to continuous operation, and can be done such that enzymatic activity is retained.

Precipitants include acids or bases (isoelectric), salts (salting-out), organic solvents, nonionic polymers, polyelectrolytes, protein-binding dyes, multivalent metal ions, and homogeneous and heterogeneous affinity ligands. Examples of the purifications obtainable using these agents can be seen in Table 1. The focus of this review will be on the selectivity conferred by those techniques towards the protein(s) they are targeting. To quantitatively characterize the selectivity of a precipitation, three different definitions are typically used: the purification factor, the separation factor, and a variant of the separation factor. Any of these definitions is suitable for comparing the selectivities of various precipitants since they can easily be calculated from experimental data. The purification factor ($\gamma$) is defined as

$$\gamma = \frac{\dfrac{[\text{target protein}]}{[\text{total protein}]} \text{in precipitate}}{\dfrac{[\text{target protein}]}{[\text{total protein}]} \text{in extract}}, \tag{1}$$

whereas the separation factor ($\alpha$) is defined as

$$\alpha = \frac{\dfrac{[\text{target protein}]}{[\text{other protein}]} \text{in precipitate}}{\dfrac{[\text{target protein}]}{[\text{other protein}]} \text{in extract}}. \tag{2}$$

The third definition is different from $\alpha$ in that the denominator is "in supernatant" instead of "in extract". This last factor is the least prevalent of the three and will not be used here, primarily since it is only a variant of $\alpha$. The purification factor is merely a representation of the increase in specific activity of the target protein compared to that in the original extract. The separation factor differs from the purification factor in that it is a measure of the increase in the ratio, not percentage, of target protein to contaminant proteins upon precipitation. A purification or separation factor greater than unity indicates an enrichment of the target protein in the precipitate. The separation factor is more

**Table 1.** Precipitation for fractionation of proteins

| Target protein(s) | Original material | % Yield | Purification Factor | Separation Factor | Precipitating agent | Ref. |
|---|---|---|---|---|---|---|
| *Isoelectric* | | | | | | |
| Prolyl-tRNA synthetase | mung bean extract | 71 | 2.6 | | Acid | [3] |
| Two proteins | soy bean extract | | | | Acid | [4] |
| Glycinin (I)[a] | | 75 | 1.9 | | | |
| β-Conglycinin (II) | | 50 | 1.6 | | | |
| *Salting-out* | | | | | | |
| Glyceral phosphate dehydrogenase | muscle extract | 74 | 2.6 | | $(NH_4)_2SO_4$ | [5] |
| Alcohol dehydrogenase | *S. cerevisiae* extract | 90 | 3.8 | | $(NH_4)_2SO_4$ | [6] |
| *Organic solvent* | | | | | | |
| Various Proteins | blood plasma | | | | EtOH | [7] |
| Fibrinogen (I) | | 22 | 2.4 | | | |
| Globulins (II & III) | | 51 | 1.8 | | | |
| Globulins (III) | | 33 | 2.1 | | | |
| Albumin (S-(II & III)) | | 87 | 1.5 | | | |
| Phytase | *A. cameus* filtrate | 73 | 1.7 | | acetone | [8] |
| *Nonionic polymer* | | | | | | |
| Alcohol oxidase | mycelial extract fraction | 77 | 1.7 | | PEG | [9] |
| α-glucosidase | *S. Carlsbergensis* extract | 70 | 5.4 | | PEG | [10] |
| α-glucosidase | *S. Carlsbergensis* extract | 50 | 6.9 | | PEG | [10] |
| Various Proteins | pig liver extract | | | | PEG | [11] |
| Glyceraldehyde-phosphatate dehydrogenase | | 70 | 4.9[b] | | | |
| Phosphoglycerate kinase | | 80 | — | | | |
| Phosphoglycermutase | | 90 | — | | | |

| | | | | | | |
|---|---|---|---|---|---|---|
| *Homogeneous macroligands* | | | | | | |
| Lactose dehydrogenase | ox heart extract | 91 | 40 | 180 | Bis-NAD | [89] |
| Avidin | pure protein^c | | | | Biocytin-dextran | [93] |
| *Heterogeneous macroligands* | | | | | | |
| Lactose Dehydrogenase | pig heart extract | 50 | 7.0 | | Con A ligand affinity | [16] |
| Trypsin | beef pancrease extract | 79 | 5.6 | 38 | Poly GAB | [37] |
| Two Proteins | mixture @ 50% each | | | 41 | Poly SAB | [91] |
| Trypsin | | 82 | 1.95 | | | |
| Chymotrypsin | | 2 | — | | | |
| Two Proteins | mixture @ 50% each | | | 25 | Poly SAB | [91] |
| Trypsin | | 74 | 1.92 | | | |
| Chymotrypsin | | 3 | | | | |
| Trypsin | porcine pancreas extract | 93 | 5.5 | | STI-chitosan | [94] |
| Wheat germ agglutinin | wheat germ extract | 70^d | 11 | e | Chitosan | [95] |
| Wheat germ agglutinin | wheat germ extract | 100^d | 11 | e | Chitosan | [95] |
| Protein A | Staph. aureus extract | 91^d | 32 | 67 | IgG-AS | [96] |
| Immunoglobulin | rabbit serum | | — | 90% pure | Protein A ligand | [97] |
| Avidin | egg white | 80 | 12 | | DMPE-B | [98] |
| *Protein-Binding dyes* | | | | | | |
| Lactose dehydrogenase | rabbit muscle extract | 97 | 6.1 | e | Triazine dye | [103] |
| Lactose dehydrogenase | rabbit muscle extract | 60 | 21 | e | Triazine dye | [104] |
| Two Proteins | S. cerevisiae extract | | | | PEG-triazine | [105] |
| Glucose-6-phosphate dehydrogase | | 93 | 3.5 | | | |
| 3-phosphoglycerate kinase | | 86 | 3.0 | | | |
| *Metal ion affinity* | | | | | | |
| Human hemoglobin | pure enzymes^c | | | | Cu(II) & Bis-Cu(II) chelates | [38, 39] |
| Whale myoglobin | | | | | | |

(Cont. of p. 164)

**Table 1** (continued)

| Target protein(s) | Original material | % Yield | Purification Factor | Separation Factor | Precipitating agent | Ref. |
|---|---|---|---|---|---|---|
| *Polyelectrolytes* | | | | | | |
| Lysozyme | egg white | 92 | 9.0 | 10.6 | PAA | [24] |
| Various Proteins | artificial mixture @ 25% | | | | PAA | [24] |
| Lysozyme (I) | | 45 | 2.5 | | | |
| Lysozyme (I & II) | | 80 | 2.4 | | | |
| Protease (IV) | | 38 | 2.2 | | | |
| Protease (IV & V) | | 59 | 2.0 | | | |
| β-galactosidase (S-V) | | 51 | 2.3 | | | |
| Two Proteins | artificial mixture @ 50% | | | | CMC | [27] |
| Lysozyme | | 100 | 1.84 | 11.2 | | |
| Ovalbumin | | 9.0 | — | — | | |
| Lysozyme | egg white | 92 | 23 | | PAA | [52] |
| Asp-11 fusion β-galactosidase | E. coli extract | 98 | — | 7.9 | PEI | [69] |
| Fugal lactase | A. awormi permeate | 85 | 14 | — | WBA | [108] |
| Alkaline protease | β-subtilis permeate | 85 | 20 | — | WPA | [108] |
| RNA polymerase I | E. coli extract | 89 | 29 | | PEI | [124] |
| RNA polymerase II | wheat germ extract | 100 | 28 | 28.2 | PEI | [125] |

[a] Roman numerals indicate precipitate fraction. "S" preface indicates supernatant
[b] Estimate based on assuming the same activity units/mass for all three enzymes
[c] No mixtures of proteins were tested
[d] Low yield due to impurities in chitosan. Yield improved to 100% after chitosan subjected to gel filtration twice
[e] A homogeneous product was obtained. The separation factor approaches infinity by definition

sensitive to the initial percentage of the target protein in relation to other proteins, as well as to the final purity achieved by the precipitation. For instance, in Table 1, the data for CMC precipitation of lysozyme from a mixture containing an equal amount of ovalbumin shows $\gamma = 1.84$ and $\alpha = 11.2$. For this lysozyme-rich starting material, $\gamma$ does not give a sensitive indication of the excellent selectivity achieved (perfect separation would give $\gamma = 2$). However, separation factors cannot always be calculated from literature sources, since data are often given only in terms of specific activity of the target protein. When available, both purification and separation factors will be used here.

Several factors should be taken into consideration when looking at the precipitation results in Table 1. First, the starting materials for the different precipitations varied widely. They include such solutions as artificial mixtures of proteins, bacterial and yeast homogenates, and filtrates of yeast cultures. Crude extracts have been centrifuged to remove cell debris, and may have undergone further processing before use in precipitation. The initial and final percentage of the target protein in relation to the total protein vary from less than 0.1% to greater than 40%. Second, neither $\alpha$ or $\gamma$ takes into account the large concentration of the product which occurs through precipitation, nor does either account for the separation from nucleic acids, lipids, etc., whose concentrations also vary widely in the initial mixtures. Both are only measures of the capability to perform the usually more difficult task of fractionating the proteins. Third, factors such as precipitation kinetics and entrainment may well have affected these results. In other words, the comparisons are not those of equilibrium processes. Such factors can lead to a "black art" perception of precipitation which deters its use. Finally, most of the data were published because by some measure the separation was successful. Only a few of the papers were reporting both successes and failures of the method.

For the purpose of narrowing this review to "selective" precipitation we will focus on only those precipitation agents demonstrating purification factors greater than 5 from crude extracts. A look at Table 1 shows that they include protein-binding dyes, macroligands, and polyelectrolytes. Despite the lack of data on metal ions, these will be included as well. Examples of several "non-selective precipitants" have been included on Table 1 for comparison [3–11]. More information on the other agents may be found elsewhere [12–14]. Protein-binding dyes and metal ions have been classified as macroligands [15], yet are considered as separate entities here due to their comparative chemical and biological stability, relative low cost, and differences in precipitating mechanisms.

Each of these "selective" methods shares the characteristic that precipitant and protein bind and it is the resulting complex that precipitates. Dye-binding associations are attributed both to electrostatic associations between charged dyes and charged regions of the protein or to the binding of a dye to a cofactor or substrate binding site on the protein. To be used as a precipitant, a dye must be capable of forming at least two associations per dye molecule. Metal ion affinity precipitation takes advantage of the interaction between divalent metal ions, such as $Co^{2+}$, $Cu^{2+}$, $Zn^{2+}$, and $Ni^{2+}$, and strong electron donor groups on the proteins. Macroligand affinity precipitation differs in that a biospecific ligand interacts with the target protein at its binding site(s). Polyelectrolyte precipitation is based on the ionic interactions between the polyelectrolyte and the proteins.

These precipitants all have their counterparts in adsorption and chromatography. Chromatographic techniques have, however, several limitations [16]: (a) resistance to mass transfer through diffusional limitations and steric hindrance in the association step lead to slow binding and low available capacity; (b) treatment of viscous or particulate matter can cause plugging and results in high pressure drops which limit flow rates; and (c) the limits of scaling up due to bead deformation with higher pressure drops. Precipitation techniques overcome these limitations and furthermore have the advantages of high throughput and continuous operation [15]. These advantages must be weighed against chromatography's advantages of multiple stages and insoluble separating agent. A comparison of purification factors obtainable by different separation methods was made recently by Bonnerjea et al. [2]. An expanded version of their findings can be seen in Fig. 1. Their survey showed the typically low purification factors expected for precipitation. However, ammonium sulphate precipitation predominated in that survey. The higher purification factors seen with "selective precipitation" represents the data from Table 1. While they still fall short of affinity adsorption, the values do approach those of other chromatographic methods. For perspective we have also included overall purification factors for 2–4 stages of aqueous phase partitioning representing 14 cases reported by Kroner et al. [17]. One should also keep in mind that as an early step precipitation is typically used in the presence of a greater number of interfering substances than would have been present in the chromatographic steps.

In choosing a precipitation strategy, factors other than the choice of precipitant must be taken into consideration. Strategies to enhance precipitation

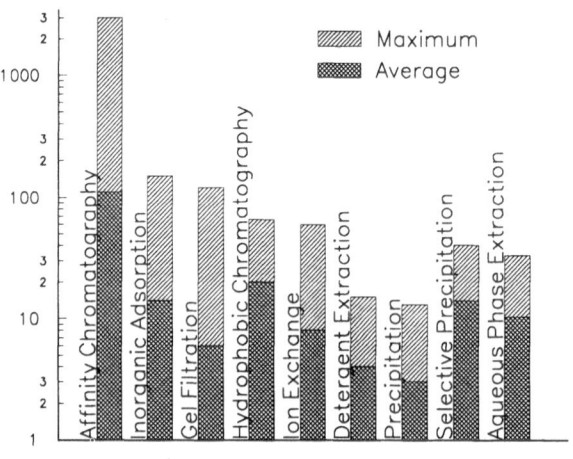

Fig. 1. Comparison of average and maximum purification factors obtainable by various separation methods (modified from Bonnerjea et al.[2])

include the manner in which the precipitant is added to the process stream and the environment in which precipitation occurs. Genetic engineering can be used to enhance the separation of the targeted protein through the fusion of peptides which confer characteristics enabling easier separation. And finally, the isolation of the product after precipitation has to be taken into account. These factors will be discussed in relation to general precipitation strategies and their application.

# 2 Choice of Precipitant

## 2.1 Binding and Selectivity

The binding between the precipitant and protein determines the selectivity of these methods. Higher specificity in binding results in greater purification of the desired product. The choice of precipitant must therefore be made so as to take advantage of any specific binding characteristics of the targeted protein.

### 2.1.1 Affinity Interactions

Affinity macroligands and protein-binding dyes selectively bind proteins through strong biospecific associations between the protein and its ligands [18–20]. The types of ligands available include substrates, coenzymes, immun-oligands, etc. The strength of binding between the protein and the ligand is typically very strong, with $K_m$ values typically ranging from $10^{-4}$ to $10^{-10}$ [21]. Binding is relatively insensitive to solution conditions such as pH, ionic strength and protein concentration, as long as the structural integrity of the protein is not impaired. The presence of competitive ligands can greatly decrease or reverse binding.

Divalent metal-ion precipitants specifically bind to surface-accessible strong electron donor groups on the protein. Of the various amino acids comprising proteins, histidine, cysteine, tryptophan, and arginine possess such groups [22]. Fractionation results from the variation in distribution of such groups on the surface of the protein.

### 2.1.2 Polyelectrolytes

The term polyelectrolyte will be used to indicate water-soluble polymers with a regular distribution of ionizable groups resulting in an expanded random coil conformation. The expansion of coil dimensions depends on chain flexibility and charge density; for weakly ionizable groups, the latter will change with degree of dissociation and hence the pH. Examples of polyanions are poly(acrylic acid) and carboxymethyl cellulose; polyethyleneimine is a poly-cation. Such molecules can complex with oppositely charged molecules to form species no longer stabilized by charge repulsion, solvation, etc. Precipitation is the result. Since polyelectrolytes are also used as flocculants (via surface-binding to aggregate particulates) in removal of suspended solids, their presence can additionally affect the character of the precipitate by this role [23].

The binding of a protein to a polyelectrolyte is believed to be dependent on electrostatic, hydrophobic, and hydrogen bonding interactions between the protein and polymer [24–26]. Electrostatic interactions are presumed to be the dominant forces. The presumption is supported by the facts that only those proteins possessing a charge opposite to the polyelectrolyte are precipitated, that highly charged proteins are selectively precipitated, and that the extent of precipitation decreases as the ionic strength is increased [27]. Further support comes from the observation that the complexation of proteins with strongly ionizable polyelectrolytes follows a stoichiometric relationship when performed under conditions in which the acidic or basic groups of the protein are completely dissociated [28, 29].

For guidance in use of such precipitants, it is worth considering these materials as soluble ion exchange resins (since such materials are essentially just crosslinked polyelectrolytes). On that basis, precipitation would be expected to be most effective at low ionic strength and at a pH where protein and polyelectrolyte have opposite charge. In this case it is the complex, including the precipitating agent, that precipitates. Hence the stoichiometry of the association will play a major role in determining required polyelectrolyte levels. The greater the polymer charge density, the greater the capacity for protein binding. Experimental studies have demonstrated the role of charge equivalence [28–30].

In addition to considering the binding capacity, one must consider the strength of the binding. This second factor will govern efficiency of removal and the selectivity of removal in mixtures. The analysis of Morrow, Carbonell and McKoy [31] for partitioning of proteins on ion exchangers gives some guidance for the basis of separation. Their results are based on a consideration of electrostatic and hydrophobic forces. The result is a partition coefficient dependent on the sum of those two interactions

$$\log\left[\frac{K_c}{K}\right] = \frac{-2AZ_pZ_bI^{1/2}}{1 + BI^{1/2}} + CI \tag{3}$$

where $K_c$ is the concentration equilibrium dissociation constant for a complex of protein and binding sites, K is the true (activity) equilibrium constant, $Z_p$ is the charge on the protein, $Z_b$ is the charge on the binding site, I is the ionic strength, and the remainder are empirical constants. The two terms on the right account for electrostatic and hydrophobic interactions. Here the protein will bind to the polyelectrolyte.

Equation 3 indicates that the net charge on the protein will be a primary basis on which separation is made. A polyelectrolyte with high charge density may ensure that this is the basis of separation by affording little opportunity for the hydrophobic interactions to become significant. Here one would want to operate away from the isoelectric point of the protein. The equation also provides for disruption of binding by increasing ionic strength.

A model specifically addressing polyelectrolyte precipitation has been developed by Clark and Glatz [32]. Their model assumes multi-site, cooperative binding. Cooperative binding accounts for the influence of bound ligands on subsequent binding. The enhancement or deterrent of further binding is termed positive or negative cooperativity, respectively [32–34]. Cooperativity is an effect often observed in biological systems. A modification of the Debye-Huckel theory is used in the theory to account for the electrostatic effects responsible for negatively cooperative binding. The model showed the effects of protein charge and ionic strength on the precipitation of ovalbumin and lysozyme with CMC.

## 2.2 Solubility and Precipitation

A typical globular protein presents to the solvent a surface consisting of positive- and negative- charged regions, polar, yet uncharged, hydrophobic regions, and nonpolar, hydrophobic regions. Proteins are of high molecular weight (typically 10,000 to 500,000 Da), yet compact, particularly in comparison with other high molecular weight components such as nucleic acids and polysaccharides. Different proteins exhibit different relative proportions of the various surface types. The complex interactions between the protein surface and surrounding solvent determine the solubility. The protein remains in solution when it is thermodynamically more favourable to be surrounded by solvent than it is to be aggregated with other protein molecules in a solid phase.

Of the specific precipitants discussed here, all except heterogeneous macroligands result in the formation of an insoluble complex upon mixing with the protein. This proceeds via the formation of an insoluble complex upon binding of the protein and the precipitant, rapid formation of a solid phase in the form of submicron primary particles, followed by larger-scale aggregation through the shear driven collision of primary particles and/or small aggregates with growing aggregates. Primary particle size typically increases with protein concentration [35]; aggregate size increases with protein concentration subject to the limits of breakage.

### 2.2.1 Macroligands

The various types of affinity precipitants can be classified as homogeneous and heterogeneous macroligands [15]. Homogeneous macroligands are ligands bound by a linker to other ligands or several bound to a macromolecule [18]. Binding of the macroligand to the protein brings about both the selective fractionation of the protein, as well as the formation of large, crosslinked aggregates. The aggregates grow until they become so large that they are no longer soluble and precipitate. For the crosslinking to occur, the target protein must bind at least two ligands (i.e. bifunctional).

Figure 2 depicts the mechanism of precipitation for homogeneous macro-ligands. The mechanism also applies to other bifunctional affinity precipitants, which include metal ions and protein-binding dyes. The crosslinking of the ligands and the proteins leads to the formation of a large, insoluble complex. An early example of bifunctional affinity precipitation used bis-NAD derivatives to precipitate dehydrogenase enzymes [18]. The NAD ligands were attached to the ends of a spacer molecule (linker), thus forming the macroligand.

The ligands of heterogeneous macroligands are responsible only for the selective fractionation upon binding. The nature of the carrier molecule deter-mines the conditions under which the macroligand-protein complex will pre-cipitate [37]. Typically, precipitation is effected through changes in the pH, ionic strength, or temperature of the solvent. The addition of a secondary precipitant which selectively crosslinks the macroligand-protein complex is another option. In this case, the result is the formation of a large, insoluble complex similar to that depicted in Fig. 2.

### 2.2.2 Protein-Binding Dyes

The formation of precipitates upon the complexation of protein-binding dyes with proteins is very similar in manner to that of homogeneous bifunctional macroligands. The protein-binding dye also functions as a bifunctional pre-cipitant, complexing with two other proteins. The result is the formation of a large, crosslinked structure as is depicted in Fig. 2.

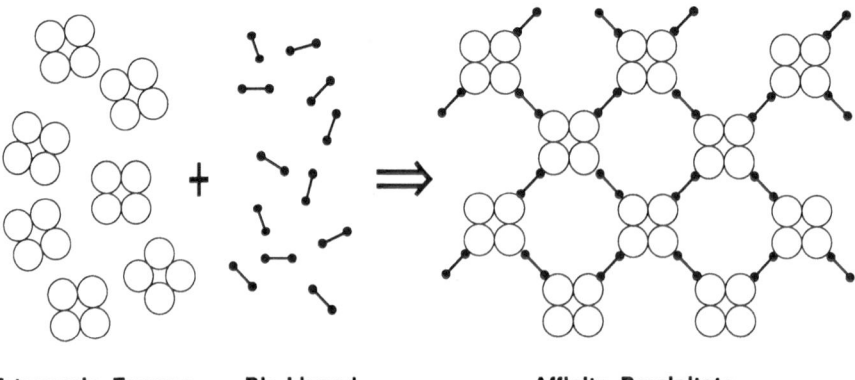

**Tetrameric Enzyme      Bis-Ligand                  Affinity Precipitate**

**Fig. 2.** Affinity precipitation using homogeneous bifunctional macroligands (after Flygare et al.[36])

## 2.2.3 Metal Ions

The complexation between divalent metal ions and the strong electron donor groups of proteins results in the formation of large, crosslinked aggregates (Fig. 2) [38, 39]. When the aggregates have grown sufficiently large, they become insoluble and precipitate. Because one ion can chelate with multiple donor groups, the precipitant can be in the form of either the simple ion or that of a bis-chelate. A bis-chelate consists of two metal ions attached to the ends of a spacer molecule.

## 2.2.4 Polyelectrolytes

The binding mechanism for the final aggregation of primary particles into flocs has been variously attributed to the mechanisms of patching [40–44], charge neutralization [45, 46], or bridging [47–49]. Figure 3 shows a schematic representation portraying these steps for the case of polyelectrolyte precipitation. The patching model assumes that parts of the primary particles remain charged upon formation. The positive and negative patches on the surfaces of the primary particles can then strongly interact to form flocs. In the charge neutralization model, the net charge on the primary particle is assumed to be nearly or completely neutralized, thus reducing protein solubility as well as the electrical repulsion between protein molecules. Polymer attachment between primary particles and aggregates to form flocs is the basis of the bridging model.

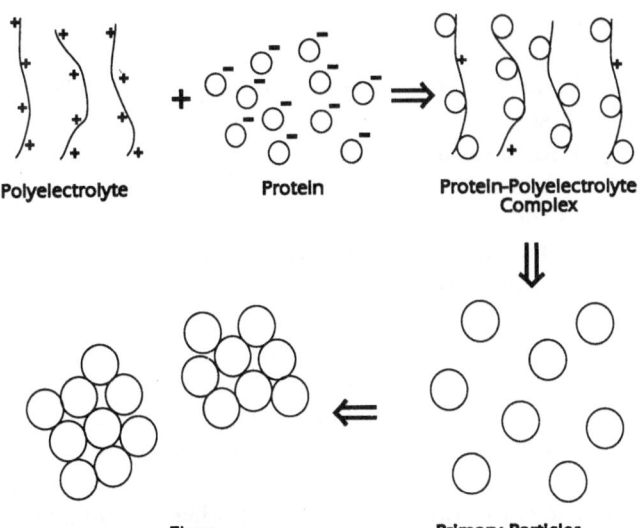

**Fig. 3.** Schematic representation of the precipitation process as proposed by Clark and Glatz[43]

# 3 Considerations Other than Selectivity

Clearly, the proper choice of precipitating agent is important to achieving high selectivity. Whatever the choice, there remain several other strategies with which selectivity may simultaneously be pursued. One must tend to the physical aspects of precipitator operation. One can consider the influence of upstream operations and potential pretreatments on the composition of the protein-containing stream. Going even further upstream, one can consider genetic engineering as a means of modifying the protein to simplify the separation task. And finally, since one has added a material separating agent, one is confronted with the task of removing the precipitant.

## 3.1 Physical Aspects of Precipitation

Protein solubility generally depends not only on concentrations of the protein and precipitant, but also on such factors as pH, ionic strength, temperature, and the concentrations of other components of the mixture. Even accounting for all of these factors would not be sufficient to determine behaviour, however, as the results frequently depend on more than thermodynamics. Of importance are such considerations as the inlet concentration and rate of addition of the precipitant, the mechanism of contacting, the duration and level of mixing, and the final recovery of the precipitate.

### 3.1.1 Addition of Precipitant

Though not among our list of "selective" methods, salting out provides an excellent example of the potentially non-equilibrium nature of solubility behavior. Figure 4 shows a series of salting out curves for fumarase obtained by adding ammonium sulphate in a variety of ways [50]. Clearly different results are obtained when the salt is added as a solid rather than as a saturated liquid. Yet another factor is seen to be important in the design of the precipitator.

The difficulty revealed in this example is that the kinetics of reagent dispersion are not sufficiently rapid to ensure that all regions of the vessel are of uniform concentration during the time that the precipitate is forming. The penalty is over- and co-precipitation. Various schemes have been proposed for achieving homogeneous nucleation of precipitates. Long ago, Meekin [51] added ethanol by dialysis and this was repeated for acid addition in isoelectric precipitation by Fisher and Glatz [52]. Other strategies include the circulation of a protein stream from a reservoir through a flow loop where precipitant is gradually introduced at a static mixer followed by recycle of the stream to the reservoir [53]. Workers at Merck [54] have used a continuous variation of this,

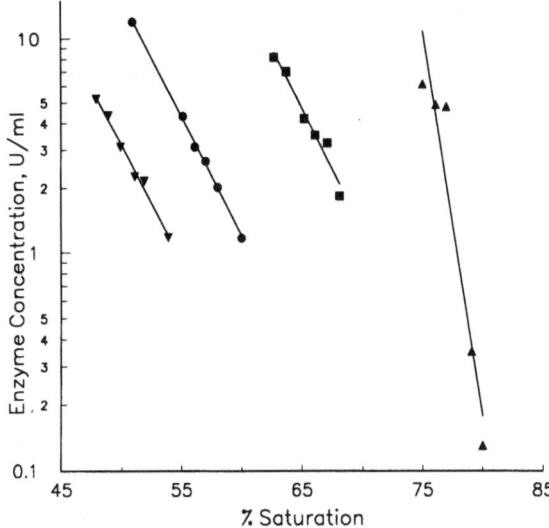

**Fig. 4.** Effect of contacting procedure on final equilibrium precipitation behavior of fumarase. ▼, ●-batch contacting; ■, ▲-continuous contacting, 16.9 minutes total residence time; ▼-ammonium sulfate solid; ●, ■, ▲-saturated ammonium sulfate solution. Other conditions were: ▼, ■, ▲-pH 5.9, 40 mg ml$^{-1}$ initial protein, 8 °C; ●-pH 5.7, 35 mg ml$^{-1}$ initial protein, 6 °C (from Foster et al.[50]) Reprinted by permission of John Wiley & Sons

joining antibiotic and precipitant in a jet mixer. In both cases, more consistent results and/or a better quality precipitate has been reported.

### 3.1.2 Mixing

The final state of the precipitate is dependent on the duration and level of mixing to which the suspension is exposed. Foster, Dunnill, and Lilly [50] reported on the changes in precipitate composition that resulted from up to several hours of aging of ammonium sulphate precipitates of yeast enzymes. A series of papers on changes in the physical characteristics of protein precipitates with exposure to shear has been reviewed by Bell, Hoare, and Dunnill [55]. Aging increases the strength of the precipitate (as evidenced by resistance to high shear breakup) up to a Camp number (the dimensionless product of shear rate and mixing time) of $10^5$ [56]. The increased strength results from restructuring of the floc and the same restructuring may affect the amount of entrained solution and behavior during solid/liquid separation steps.

The initial formation of the floc structure is also controlled by mixing as floc growth is the result of hydrodynamically driven collisions of primary particles and small aggregates with the growing flocs [57]. For most precipitations the initial formation of primary particles occurs on the order of seconds and even floc growth is largely complete within tens of seconds. In contrast, completion of precipitation using a homogeneous macroligand to recover a dilute protein from a complex mixture has been reported to take several hours [58]. And while polyelectrolyte precipitation proceeds with typical rapidity at optimal pH conditions, the rate was observed to be much slower two pH units away from the optimum [59].

### 3.1.3 Recovery of the Precipitate

After the precipitate is formed, it must be separated from the liquid and the protein subsequently separated from the precipitant. To carry out the former requires an operation such as settling, flotation, centrifugation, traditional filtration, or cross-flow membrane filtration. For most of these methods, particle size becomes the determining factor in the ease and speed of the process. Hence, the formation of large flocs is desirable. For these techniques involving filtration, an additional problem is the high cake resistance of the typically compressible protein precipitates. The use of filter aids and/or cross-flow operation are generally necessary to overcome this problem. Further information on the nature of aggregate strength and its consequences can be found elsewhere [60].

## 3.2 Nature of the Extract

Obtaining a pure product is particularly complicated when one considers the fact that cells consist of a complex mixture of many different components, including cell walls (lipid bilayers), ribosomes, nucleic acids, and proteins. To complicate matters further, even in such simple cells as bacteria there are on the order of 1000 different proteins.

The demands on product purification are dependent on the bioconversion process. If the product is secreted by the microorganism, the product will be fairly dilute in the broth, yet relatively pure. On the other hand, if the product is contained within the microorganism, either in the periplasmic space or in the cytosol, the cells must be harvested and subsequently disrupted to recover the product, which will be concentrated, yet relatively impure. Dilution often improves product recovery from streams (such as cheese whey) which possess prohibiting concentrations of interfering components (such as salts) [61]. To avoid unnecessary product losses, conditions during the bioconversion and purification must be adjusted so as to avoid denaturation or proteolytic degradation of the product.

Components such as cell debris and nucleic acids can interfere with, and must often be removed prior to, protein precipitation [62, 63]. The removal of cell debris prior to the precipitation of crude extracts is necessary to avoid contamination of the precipitate, and thus the product, by the cell debris [64, 65]. Polyelectrolytes have been used to combine the removal of nucleic acids and cell debris in one step [66–68]. Evidence as to the improvement in selectivity upon removal of nucleic acids can be found in the precipitation of genetically modified β-galactosidase with PEI [69].

## 3.3 Impact of Genetic Engineering

Genetic engineering can be used to produce greater product yields and to enhance the separation characteristics of targeted proteins. Through increased expression levels (up to 45% of total protein in the cell [70, 71] and choice of microorganism, recombinant organisms can now create mass quantities of products which were once very difficult to extract, and thus very rare and expensive [65, 72].

Enhancing the separation of a protein has typically been done by genetically fusing a polypeptide conferring a basis for recovery to the terminus of the desired protein. Such genetic fusions are referred to as purification fusions or "tails". The tails have included charged amino acids for recovery by ion exchange, multiple histidines for recovery by metal ion affinity, and whole proteins for recovery by ligand affinity [73–78]. Many fusion tails do not interfere with the activity of the enzyme and may even offer protection from proteolytic degradation [79–84]. Thus far the tails have primarily served for adsorptive binding, but the technology is readily adaptable to application in precipitation.

For example, ion exchange chromatography has previously been shown to enhance the recovery of a genetically modified small protein, β-urogastrone, containing positively charged fusion tails [85, 86]. Similarly, negatively charged aspartic acid tailed β-galactosidase fusions showed improved separation behavior relative to wild-type β-galactosidase on an ion exchange column [87, 88]. Precipitation of β-galactosidase fused with a tail of 11 aspartic acid residues with polyethyleneimine at a precipitant:protein ratio of $0.01 \text{ gg}^{-1}$ resulted in a separation factor of 7.9. This contrasts with the unmodified protein which was not enriched in the precipitate formed under the same conditions [69]. The fusion did not interfere with the enzymatic activity. Given the size of and large number of charged groups on β-galactosidase, application of this strategy to smaller proteins may well give even better results.


## 3.4 Removal of Precipitant

The separation of the protein from the precipitant is necessary most of the time. Economics often require the recovery and reuse of the precipitant. Also of concern is whether the precipitants are acceptable agents for bioprocessing. Some precipitating agent are not approved for use with food or pharmaceutical products. Other agents may interfere with subsequent processing steps. Typically, the precipitate is washed to remove any non-specifically included substances. It is then resuspended in a buffer in which the precipitating agent and the protein dissociate. This buffer is preferably one which can be directly used in subsequent processing steps. In the case of heterogeneous macroligand affinity precipitation, the macroligand can be designed so that it remains insoluble upon dissociation from the protein. For cases in which both the precipitant and protein remain soluble, separation can be achieved using ultrafiltration or chromatographic methods. If the protein and precipitant are of substantially different size, ultrafiltration can be used to achieve quick separation. Ultrafiltration has the advantage of continuous operation. Chromatographic techniques offer a variety of bases upon which separation can be accomplished such as size, charge, and hydrophobicity.

# 4 Use of Selective Precipitants

In addition to the separation factor, the following must be considered in choosing a precipitant for a specific separation task:

1. Are traces of the soluble precipitant acceptable?
2. How can the soluble precipitant be separated from the target protein?
3. Is the precipitant stable in the extract?
4. How much of the precipitant is needed and how critical is the dosage?
5. Is the process sensitive to the values of pH, ionic strength, temperature, etc., and if so, to what extent?

Each of the selective precipitation methods has different answers to these questions. The following sections discuss these and other important aspects for each precipitating agent.

## 4.1 Affinity Macroligands

The basis of affinity macroligand precipitants lies in the attachment of a ligand to a soluble carrier. Binding of this macroligand to the target protein results in the formation of a complex which can be precipitated out of solution. Affinity precipitation can be divided into two categories corresponding to the nature of the precipitant: homogeneous macroligands and heterogeneous macroligands.

### 4.1.1 Homogeneous Macroligands

Homogeneous bifunctional macroligands consist of two identical affinity ligands connected by a spacer; also termed bis-ligands. The first extensive studies of affinity precipitation used bis-NAD to precipitate multimeric dehydrogenases [18, 36, 58, 89, 90]. For bis-ligand affinity precipitation to work, the following requirements have to be met [90]:

1. The enzyme must have more than one binding affinity site.
2. The bifunctional ligand must have a strong affinity for the enzyme.
3. The spacer which binds the ligands must be sufficient length to bridge the distance between the binding sites on the enzymes, yet not so long as to bind two sites on the same protein.

The precipitant has the advantage of biospecificity, which can lead to very high purifications. The conditions of precipitation are typically mild enough that neither protein nor ligand is denatured [91]. Disadvantages include the high cost and susceptibility to proteolytic degradation of homogeneous macroligands. Recovery of the macroligand is therefore very important.

Research using bis-biotinyl compounds to selectively and strongly bind avidin displayed a marked effect of linker length upon the final structure of the complex [92]. Only one of the two biotin residues separated by polymethylenediamine linkers of less than 14 Å were able to bind to an avidin molecule, whereas linker lengths exceeding 38 Å resulted in the reagent binding two subunits of the same avidin molecule. Intermediate linker lengths were found to form intermolecular complexes. Larsson and Mosbach studied the role of spacer length using bis-NAD with spacer lengths of 7, 17, and 32 Å [18]. A spacer length of 17 Å (using $N^2$, $N^{2'}$-adipodihydrazido-bis ($N^6$-carbonylmethyl-NAD)) was found to be optimal in precipitations with various dehydrogenases. The selectivity and effective strength of binding of the bis-NAD system were increased through ternary complex formation by adding competitive inhibitors to the solution. For example, the addition of the competitive inhibitors pyruvate or oxalate to precipitations involving the tetrameric enzyme lactate dehydrogenase resulted in the formation of strong ternary complexes [89].

The yield of the precipitation using bis-NAD macroligands was found to be dependent on the ratio of NAD ligands to enzyme subunits. An optimum would be expected at ratios near unity. At ratios higher than unity, a decrease in yield is expected since not all ends of the bis-ligands will be able to complex with enzyme subunits. At ratios lower than unity, not all enzymes will be bound. Experimentation confirmed an optimum near unity for precipitations of LDH at a ratio of 1.25 NAD ligands:LDH subunits. Precipitations with oligomers possessing a greater number of subunits are affected to a lesser extent by deviations from a ratio of unity. The hexamer glutamate dehydrogenase yielded almost quantitative precipitation between ratios of 0.3 and 10. A problem which may be encountered with this method is the formation of linear polymers or dimeric complexes. The latter problem was encountered when attempting to precipitate liver alcohol dehydrogenase [89]. Soluble complexes were formed which consisted of two enzymes bound by two bis-NAD molecules.

In order to observe selectivity in a mixture of enzymes, precipitation of LDH was performed on ox heart crude extract. The precipitation resulted in a purification factor of 40 and a yield of 91% for greater than 95% purity [89]. However, the time required for complete precipitation was relatively long with a minimum of two hours. Dissolution of the complex after centrifugation was accomplished by the addition of the competitive ligand NADH.

A more recent development in using homogeneous macroligands for affinity precipitation is the binding of multiple ligands to a polymer [93]. Preliminary results using biocytin bound to dextran for the precipitation of avidin appear to be promising. Up to 90% of the avidin in solution could be bound under optimal conditions. The same conditions yielded only 3% precipitation when using lysozyme as a control. Optimum precipitation was obtained using low ionic strength and highly substituted, low molecular weight dextran.

### 4.1.2 Heterogeneous Macroligands

The use of heterogeneous macroligands has several advantages over the use of bifunctional affinity ligands. First, rather than a sharp increase in yield at the required stoichiometric ratio of ligand to protein, the yield steadily increases with the amount of heterogeneous macroligand added, making it easier to adapt the precipitation to changes in protein concentration. Second, precipitation is not coincident with complex formation. The mechanism by which precipitation occurs can be chosen independently from the binding step. The only restriction is that the conditions for precipitation do not also dissociate the protein-ligand complex. Subsequent separation of protein and macroligand is easiest if there are conditions where only the protein can be extracted from the precipitate. Thirdly, heterologous precipitants are not limited to multimeric proteins. However, the ligand moieties still suffer from their susceptibility to proteolytic degradation as well as their high cost. Protein activity retention has been good for the cases reviewed here.

Schneider et al. [37] first demonstrated this technique. They used a terpolymer of $N$-acrylol-$p$-aminobenzamidine, acrylamide, and $N$-acryloyl-$p$-aminobenzoic acid which is soluble at neutral pH but insoluble in the acid form at low pH. The $p$-aminobenzamidine moiety acts as the affinity ligand for trypsin. After binding, precipitation was induced by lowering the pH to 4.0. Dissociation of the complex was accomplished by further lowering the pH to 2.0. Application of the procedure to beef pancreas extract resulted in a trypsin yield of 79% and a separation factor of 38 for 90% purity. The remaining protein in the precipitate consisted essentially of the very similar protein chymotrypsin, yet only 6.3% of the original chymotrypsin was present in the precipitate. After re-using the macroligand up to eight times, the separation factor decreased only to 31, giving 84% pure trypsin at a yield of 76%. The average loss of the macroligand was given as 1% per cycle; 93% of the macroligand remained after 8 cycles.

Other pH-dependent heterogeneous macroligands include those based on chitosan [94, 95] (used as the backbone in the recovery of trypsin and WGA) and hydroxypropyl methylcellulose acetate succinate (AS) [96] (used for recovery of protein A). Chitosan, a partly deacetylated chitin (obtainable from shrimp and crab shells), is rich in the polycationic repeat unit $N$-acetyl-D-glucosamine and is insoluble above pH 6.5. The macroligand allowed for dissolution and protein dissociation at pH 2.5. Trypsin was separated from the soluble chitosan macroligand by gel permeation chromatography [94]. In the affinity precipitation of WGA with chitosan [95], dispersed gas flotation was used in place of centrifugation for precipitate isolation. AS, insoluble below pH 4.5, was coupled to IgG to bind protein A [96]. After precipitation at pH 4.5, protein A was able to be extracted from the precipitate at pH 2.5. Recyclability of the macroligand was demonstrated over four cycles with an average yield of 91% and a separation factor of 67.

Temperature and ionic strength have been used for solubility control with NIPAM-GMA (a copolymer of $N$-isopropyl acrylamide (NIPAM) and glycidyl methacrylate (GMA)) as the carrier molecule for the trypsin-binding ligand PAB. Raising the temperature above 34 °C gave macroligand and trypsin recoveries of 95% and 82%, respectively, from trypsin/chymotrypsin mixtures, with only 2% of the chymotrypsin coprecipitating [91]. This backbone was also used as a carrier for IgG to recover protein A conjugates.

In addition to using pH, temperature, or ionic strength variations to induce precipitation, a method has been recently introduced which uses a biospecific crosslinking agent to induce precipitation [16]. The multivalent lectin concan-avalin A (Con A) was used as the agent to biospecifically crosslink the Blue Dextran macroligand/LDH complex by binding glucose residues and thereby effect precipitation. The LDH was bound to the Cibacron blue residues. While the precipitation was independent of the ratio of Blue Dextran to LDH, the ratio of Con A to Blue Dextran had to be optimized. Recovery of the target enzyme from the complex is complicated by entrapment of the enzyme in the complex and by the presence of Con A. These workers removed the Con A by binding to DEAE-Trisacryl gel and the blue dextran by gel filtration. A similar approach was used to recover IgG from serum by binding to protein A immobilized on galactomannan followed by precipitation through non-covalently crosslinking with $KBO_4$ [97]. Dissociation of IgG was accomplished by adding KSCN. Subsequently, the macroligand-borax complex was dissociated by lowering the pH.

A novel approach has recently been demonstrated using affinity surfactants to specifically precipitate multimeric proteins [98]. The macroligand consists of an affinity ligand which has been covalently attached to the polar head group of the surfactant. The study focussed on the use of dimyristoylphosphatidyleth-anolamine-biotin (DMPE-B) to selectively precipitate the egg white protein avidin. Precipitation was thought to result from the binding of a single avidin to four macroligands. The hydrophobic tails then interact to form a network similar to that proposed for homogeneous bis-ligands (see Fig. 2). Precipitations of CMC-pretreated (to remove hydrophobic and aggregating impurities) hen egg whites resulted in 91% of the avidin being removed. Greater than 80% of the lysozyme activity was retained after centrifugation, resolubilization, dis-sociation of the complex by denaturation, ultrafiltration to remove DMPE-B, and renaturation. The corresponding separation factor achieved by the overall process is 110. Among the advantages of this process over other affinity precipitation methods are that the synthesis of the macroligand is generally simpler and cheaper, and that the phospholipid does not contain any charges which could lead to non-specific interactions.

## 4.2 Protein-Binding Dyes

Triazine dyes have become widely used in protein purification. When attached to solid supports, triazine dyes exhibit high protein-binding capacities towards some proteins and the bound protein can easily be dissociated under mild conditions [99]. The binding is thought to be largely ionic and the resulting protein-dye complex is more hydrophobic and can precipitate [19]. An example of such a dye precipitant is Rivanol, an organic cation which has been used to purify serum proteins [14].

Recently, triazine dyes which specifically bind certain classes of proteins apparently through affinity interactions have been described. Bis-ligand affinity precipitants have been constructed by covalently linking two of these dye molecules (i.e. Cibacron Blue) via a spacer molecule [100, 101]. However, attempts to use poly- (Cibacron Blue) conjugates to precipitate LDH were unsuccessful [102]. The use of protein-binding dyes rather than biological ligands for use in affinity macroligand precipitants has the advantage of lower cost and stability when exposed to crude cell lysates.

Among the proteins bound by Cibacron Blue are the NAD-dependent dehydrogenases. The dye has been derivatized to increase its specificity towards the dehydrogenases [20, 103, 104]. Earlier experiments with bis- (Cibacron Blue) derivatives showed limited selectivity towards the cofactor-dependent enzymes [100, 101]. However, a simple methoxylated derivative of the $p$-sulphonate isomer of Cibacron Blue F3G-A resulted in specific precipitation of LDH from rabbit muscle crude extract [103].

The precipitation is believed to be the result of crosslinking the LDH molecules with the dye to form a large insoluble complex. The dye acts as the functional analogue of bis-NAD, with the anthraquinone moiety serving as one binding site, the methoxytriazinyl ring and terminal $p$-aminobenzenesulfonate ring serving as the other binding site, and the central $p$-phenylenediaminesulfon-ate ring acting as the linker (see Fig. 5). Support for this mechanism comes from the fact that rapid dissociation is achieved by the addition of relatively low concentrations of competitive ligands such as NADH. Furthermore, the maxi-mum precipitation displays the optimal molar ratio of enzyme subunit:dye of 2:1, which is expected for such a mechanism. The entire cycle time for preparative precipitation, including tissue homogenization, DEAE-Sepharose pretreatment, enzyme precipitation and dissolution, and separation of the enzyme and precipitant via gel chromatography was approximately 3 hours [104].

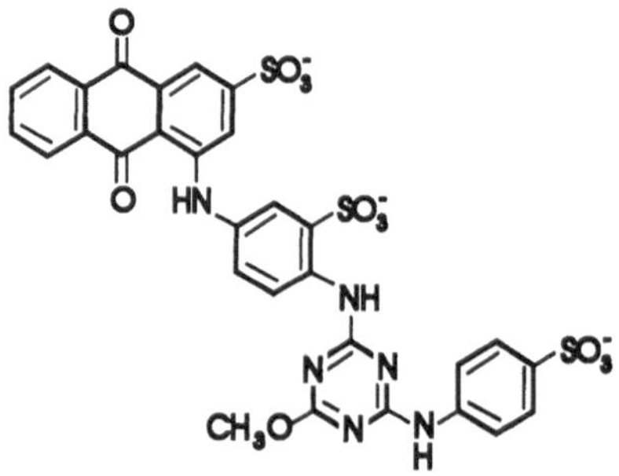

Fig. 5. Structure of the methoxylated $p$-sulphon-ate isomer of Cibacron Blue F3G-A (after Pear-son et al.[104])

"Affinity constrained precipitation" is an interesting variation in this precipitation strategy which operates by forming a soluble affinity complex with the target protein under conditions which precipitate other proteins. Johansson and Joelsson [105] used high concentrations (12.5% v/v) of PEG to precipitate undesirable proteins while a small fraction of PEG with attached dye moieties formed soluble complexes with dye-binding proteins. The method was used to purify glucose-6-phosphate dehydrogenase and 3-phosphoglycerate kinase from a crude extract of baker's yeast. The overall yield was 93% for a 3.4-fold purification.

## 4.3 Metal Ions

Experimentation using metal affinity precipitations has been limited thus far to demonstrating the use of bis-copper chelates to crosslink proteins containing multiple surface accessible histidine residues [38, 39]. No data currently exists for precipitations from crude extracts. Two different bis-chelates were used in these studies: PEG-Cu(II), composed of cupric cations chelated by molecules of iminodiacetic acid and immobilized on each end of polyethylene glycol (PEG-2000), and $Cu(II)_2EGTA$, composed of two cupric cations chelated by a molecule of ethylene glycol bis($\beta$-aminoethyl ether)$N_1N'$-tetraacetic acid (EGTA). The bis-copper chelates were shown to be effective in precipitating proteins such as human hemoglobin and sperm whale myoglobin which have multiple surface-accessible histidine residues (26 and 6 respectively). Precipitations with Cu(II)EGTA showed that human hemoglobin precipitated to 100% to a copper to surface accessible histidine ratio of unity, whereas at this concentration, sperm while myoglobin was precipitated to less than 10%. A protein which contains only one surface-accessible histidine, horse heart cytochrome c, could not be precipitated even when large quantities of bis-chelates were added. The higher molecular weight PEG-Cu(II) was shown to be a more effective precipitant on a molar basis. Precipitations carried out using an excess of bis-chelate revealed a 1:1 stoichiometric ratio of copper:surface-accessible histidine in the precipitate. Of importance in the design of such bis-chelates is that the metal ions bind strongly to the carrier so that they will not be lost under the conditions necessary to dissociate the protein from the complex.

The use of recombinant technology should lend itself readily to improving the separation characteristics of a protein via metal-ion affinity precipitations. Possibilities for application in precipitation can be seen in recent articles on the metal-ion affinity chromatography of recombinant proteins [22, 106, 107]. For example, the addition of two histidine peptides to the carboxyl end of mouse dihydrofolate reductase resulted in greatly enhanced recovery of the enzyme on an immobilized nickel column [107]. Fusions of up to six histidine residues at either end of the enzyme displayed an increasing affinity for the column with increasing number.

## 4.4 Polyelectrolytes

For polyelectrolyte precipitation, the net effect of the electrostatic repulsion between protein molecules is minimized upon complexation between the polyelectrolyte and protein [108]. Advantages of the method include high removal efficiencies and retention of enzymatic activity [24, 27–29, 109, 110]. High removal efficiencies result partly from the fact that polyelectrolytes can disrupt already existing associations, or be performed at a pH which does so. Very low amounts of polyelectrolyte (0.05–0.10% wt/vol) are required and the fractionation potential is good [13]. Since the ionic moieties on polyelectrolytes can range from strong acids to strong bases, precipitants are available over a wide pH range. Yield and floc characteristic are dependent on the polyelectrolyte/protein ratio. Reclamation and recycling of the polymers can be accomplished [111, 112]. Separation of the polyelectrolyte from the protein can be accomplished on the basis of charge, size, or solubility. Some of the polyelectrolytes have also been approved for ingestion.

The studies of polyelectrolyte precipitation include the fractionation of artificial mixtures of proteins [24, 52, 61, 113, 114], nucleic acids [63, 68, 115], the recovery of whey proteins [114, 116–118], and the fractional recovery and isolation of serum glycoproteins [119], recA protein [120], and viral proteins [115]. Factors which have been found to affect polyelectrolyte precipitation include system pH and ionic strength; the molecular weight, charge density, dosage, and type (ionic group and backbone) of the polyelectrolyte; and the size and surface characteristics of the protein.

The degree to which various proteins will interact with a polyelectrolyte under identical solution conditions depends on the surface characteristics of the individual proteins. Both the number and distribution of charged sites on the protein surface determine the strength of the protein-polyelectrolyte complex [30]. Only proteins possessing a charge opposite to that of the polyelectrolyte are precipitated, and those of higher charge density are precipitated preferentially [27]. Several authors have demonstrated this effect by precipitating artificial mixtures of proteins with polyelectrolytes. Sternberg and Hershberger [24] fractionated a mixture of four proteins with PAA, whereas Clark and Glatz [27] precipitated a binary mixture of lysozyme and ovalbumin with the polyanion CMC. In the latter experiment, complete separation between the two proteins was obtained at neutral pH values where only the lysozome possessed a net positive charge.

The selection of system pH and ionic strength are critical since polyelectrolyte precipitation is a charge-based separation. Several authors [26, 27, 30, 109, 114, 117, 118] have demonstrated that increasing the ionic strength leads to a decrease in the separation factor and higher polymer dosage requirements. Furthermore, Hill and Zadow [118] found that this effect of ionic strength on precipitation was dependent upon the polyelectrolyte used to effect precipitation. Increased ionic strength reduces the effect of the pH on precipitation behavior and can even enhance fractionation selectivity if the target protein is highly charged [27]. The optimum pH for a given process will be dependent on the particular protein [26] as well as the associated polyelectrolyte [121], since a change in pH affects the protein charge distribution and the net charge of both. These solution characteristics may also be important in that they may affect the flexibility of the polyelectrolyte and thereby influence precipitation through steric factors.

Protein recovery levels have been found to increase upon increasing charge density, molecular weight, and dosage of the polyelectrolyte [24, 26, 59, 114, 117, 122]. These same authors have emphasized the need for careful control of the polyelectrolyte dosage, which has been found to be dependent on the charge density and molecular weight of the polyelectrolyte, its degree of ionization, and on the target protein. Addition of excess of polyelectrolyte can result in reduced protein recovery, an effect which may be less severe for proteins and polyelectrolytes of high charge density [27] and polyelectrolytes of high molecular weight [59]. Hidalgo and Hansen have proposed that the loss from resolubilization is due to a redistribution of protein in the complex as more polyelectrolyte becomes available [114]. The result is a change in the stoichiometry of the complex. Support for this theory comes from measurements of the zeta potentials of aggregates made by Clark and Glatz [27]. They found that the zeta potential decreased as the polymer dosage increased, indicating that the proportion of polymer in the complex increases with increasing dosage.

Precipitations of enzymes from various crude extracts have yielded impressive results. A sampling of results for precipitations involving various polyelectrolytes, including PAA, CMC, and PEI, as well as two heteropolyacids (WBA and WPA) can be seen in Table 1. Typical yields exceed 90% for purification factors as high as 29.

An experimental protocol typical of polyelectrolyte precipitation can be seen in the precipitation of lysozyme from egg white using PAA [24]. The precipitation was carried out at room temperature and required only 5 minutes. Dissolution of the recovered precipitate was accomplished by raising the ionic strength and the pH of the system. Due to the difference in sizes between lysozyme and PAA, ultrafiltration was used to remove the polyelectrolyte and obtain a permeate containing purified lysozyme.

Another polyelectrolyte commonly used in precipitation is the cationic PEI. A problem encountered with the use of PEI to effect precipitation from cell homogenates is the associated interference from binding of nucleic acids to PEI. In fact, it has been shown that selective removal of nucleic acids from crude extracts can be accomplished by PEI precipitation without much loss of proteins if the ionic strength is above 1.0 M at neutral pH [63, 109].

A high charge density on the protein can partly offset this interference from nucleic acids. When peptides of 5 and 11 aspartic acid residues were genetically fused to the carboxyl end of β-galactosidase from E. coli, precipitation with poly (ethyleneimine) was enhanced [69, 87]. Precipitation of crude cell extracts revealed that the longer tailed enzyme could be selectively separated from solution at high yield (85%) with a separation factor of 2.2, whereas the shorter tailed version and native enzyme could not [29]. Nonetheless, a pretreatment of the extract to remove the nucleic acids lead to still higher selectivity. Additional examples of selective precipitation using PEI can be found in the review by Jendrisak [109].

Sternberg demonstrated the use of heteropolyacids to selectively precipitate proteins from culture supernatants [108]. Heteropolyacids are different in nature from the polyelectrolytes discussed above. The degree of polymerization of heteropolyacids is dependent on the pH of the solution. Polymerization occurs in acidic solutions. Heteropolyacids are formed in solutions containing molybdate or tungsten and other oxo ions (e.g. $PO_3^{3-}$, $SiO_4^{4-}$) or metal ions [123]. After recovery of the precipitate, the protein can be obtained by raising the pH whereupon the heteropolyacid depolymerizes, resulting in dissolution of the complex. The binding between the heteropolyacid and the protein was attributed primarily to ionic interactions, although a degree of affinity towards certain amino acids was found.

# 5 Summary

The use of precipitation has been shown to be an efficient method for selective separation of proteins from crude biological mixtures. The precipitants which have demonstrated highest selectivity include polyelectrolytes, metal ions, protein-binding dyes, and biospecific affinity macroligands. Affinity macroligands selectively precipitate a target protein through biospecific binding to a ligand and can be classified as either homogeneous or heterogeneous, depending on how precipitation is induced. Heterogeneous macroligands allow for the properties which effect precipitation to be chosen independently of the protein-binding properties. Affinity macroligand precipitation can be extended to a wider range of proteins through genetic engineering or purification fusions. Major disadvantages of these precipitants include biodegradation and high cost. The three other precipitants overcome these disadvantages to a great extent, yet are generally not as specific towards the targeted protein. Polyelectrolytes effect separation on the basis of charge. The precipitant is relatively cheap, stable in crude biomixtures, and can be recycled. Their use can also be extended through genetic fusions. Literature concerning the other two methods is sparse. The affinity of metal ions towards histidine residues has been demonstrated in precipitation, yet wider applicability may be found by looking at current uses in chromatography. Such observation shows that proteins can be modified by the addition of metal ion affinity fusion peptides which should enhance their separation via precipitation. Selective binding of derivatised dyes to proteins has been shown to occur through affinity interactions. The precipitant is a less expensive version of the affinity macroligands and is stable in biological mixtures. Broad applicability is hindered by the limited variety of dyes which have been found to be selective.

Factors other than the choice of precipitant must also be taken into account in designing a precipitation process for maximum selectivity. Such factors include the manner in which the precipitant is introduced into the process stream, the environment in which precipitation will occur, the solids/liquids separation after precipitation, and the subsequent separation of the targeted protein from the precipitant. The optimization of each of these factors will vary depending on the precipitant to be used, as well as on the target protein. Special attention must be given to the design so as to minimize the losses of either precipitant or protein.

Given that these factors have been properly optimized, selective precipitants can achieve selective recovery of a target protein while at the same time effecting a substantial concentration. The power of using selective precipitants to achieve protein fractionation is evidenced by experiments where homogeneous protein products have been isolated in excess of 90% overall yield from crude biological mixtures.

# 6 References

1. Böing JTP (1982) In: Reed G. (ed) Prescott and Dunn's industrial microbiology, 4th edn, Chap 15, AVI, Westport
2. Bonnerjea J, Oh S, Hoare M, Dunnill P (1986) Bio/Technology 4: 954
3. Dunnill P, Currie JA, Lilly MD (1970) Biotech Bioeng 12: 63
4. Fisher RR, Glatz CE, Murphy PA (1986) Biotech Bioeng 28: 1056
5. Bentley P, Dickinson FM, Jones IG (1973) Biochem J 135: 853
6. Richardson P, Hoare M, Dunnill P (1990) Biotech Bioeng 36: 354
7. Watt JG (1970) Vox Sang 18: 42
8. Ghareib M (1990) Acta Microbiologica Hungarica 37: 159
9. Janssen FW, Ruelius HW (1968) Biochim Biophys Acta 151: 330
10. Hönig W, Kula M-R (1976) Anal. Biochem. 72: 502
11. Kula MR, Hönig W, Foellmer H (1978) In: Sandberg HE (ed) Proc. Int'l Workshop for Protein Separation and Improvement of Blood Plasma Fractionation USDHEW Publ. No. (NIH) 78-1422, p 361, Washington, D.C.: U.S. Govt's Printing Office
12. Glatz CE (1989) Separation Processes in Biotechnology (Asenjo JA (ed)), Chap. 11, Marcel Dekker, New York
13. Scopes RK (1987) Protein purification: Principles and practice, 2nd edn, Springer, Berlin Heidelberg New York
14. Rothstein F, Rosenoer VM, Hughes WL, Current Concepts Concerning Albumin Purification (1977) In: Albumin Structure, Function and Uses (Rosenoer VM, Oratz M, Rothschild MA (eds)), Oxford: Pergamon
15. Chen J-R (1990) J Fermentation and Bioeng 70 (3): 199–209
16. Senstad C, Mattiasson B (1989) Biotechnol Appl Bioeng 11: 41
17. Kroner KH, Hustedt H, Kula M-R (1984) Process Biochem. 19(Oct): 170
18. Larsson PO, Mosbach K (1979) FEBS Lett 98: 333
19. Bertrand O, Cochet S, Kroviarski Y, Truskolaski A, Biovin P (1985) Chromatogr. 346: 11
20. Pearson JC (1987) In: Reactive Dyes in Protein and Enzyme Technology (Clonis YD, Atkinson T, Bruton CJ, Lowe CR (eds.)), p 187, New York: Stockton
21. Stryer L (1988) Biochemistry, New York: W.H. Freeman

22. Ljungquist C, Breitholtz A, Brink-Nilsson H, Moks T, Uhlen M, Nilsson B (1989) J Biochem 186: 563
23. Clark KM, Glatz CE (1986) Paper presented at the Annual Meeting of Amer Inst Chem E, Miami Beach
24. Sternberg M, Hershberger D (1974) Biochem Biophys Acta 342: 195
25. Gekko K, Noguchi H (1978) J Agric Food Chem 26: 1409
26. Zadow JG, Hill RD (1975) J Dairy Res 42: 267
27. Clark KM, Glatz CE (1990) In: Downstream Processing and Bioseparation: Recovery and Purification of Biological Products (Hamel J-FP et al (eds)), ACS Symposium Series vol 419, p 170, Washington DC: ACS
28. Kokufuta E, Takahashi K (1989) Polymer 31: 1177
29. Kokufuta E (1991) Polymer Preprints, 32(1): 604
30. Ledward DA (1978) Protein-Polysaccharide Interactions, In: Polysaccharides in Food (Blanshard JMV, Mitchell JR (eds)) p 205, London: Butterworths
31. Morrow RM, Carbonell RG, McKoy BJ (1975) Biotech Bioeng 17: 895
32. Clark KM, Glatz CE Chem Eng Sci, (accepted for publication, 1991)
33. Connors KA (1987) Binding Constants, In: Binding Constants: The Measurement of Molecular Complex Stability ((eds)), p 21, New York: John Wiley
34. Van Holde KE, Physical biochemistry (1971) In: Foundations of Modern Biochemistry Series (Hager L, Wold F (eds.)), p 57, Englewood Cliffs: Prentice-Hall
35. Nelson CD, Glatz CE (1985) Biotechnol Bioeng 27: 1434
36. Flygare S, Månnsson M-O, Larsson P-O, Mosbach K (1982) Appl Biochem Biotech 7: 59
37. Schneider M, Guillot C, Lamy B (1981) Annals NY Acad Sci 369: 257
38. Van Dam ME, Wuenschell GE, Arnold FH (1989) Biotech Appl Biochem 11: 492
39. Suh S-S, Van Dam ME, Wuenschell GE, Plunkett S, Arnold FH. (1990) In: Protein Purification: From Molecular Mechanisms to Large-Scale Processes (Ladisch MR, Willson RC, Painton CC, Builder SE (eds)), vol. 427, p 139. ACS
40. Mabire R, Audebert R, Quivorou C (1984) J Colloid Interface Sci 97: 120
41. Lindquist GM, Stratton RA (1976) J Colloid and Interface Sci 55: 45
42. Gregory J (1973) J Colloid Interface Sci 42: 448
43. Gregory J (1976) J Colloid Interface Sci 55: 35
44. Strege MA, Dubin PL, West JS, Flinta CD (1990) In: Downstream Processing and Bioseparation: Recover and Purification of Biological Products (Hamel J-FP, Hunter JB, Sikdar SK (eds)), vol. 419, p 158, Washington DC: ACS
45. Sato T, Ruch R (1980) Stabilization of Colloidal Dispersions by Polymer Adsorption, p 122. New York: Marcel Dekker
46. Gregory J (1968) Trans Faraday Soc 65: 2260
47. Hogg R (1984) J Colloid Interface Sci 102: 232
48. LaMer VK, Healy TW (1963) Rev Pure Appl Chem 13: 112
49. LaMer VK, Healy TW (1963) J Phys Chem 67: 2417
50. Foster PR, Dunnill P, Lilly MD (1976) Biotech Bioeng 18: 545
51. Meekin TC (1939) JACS 61: 2284
52. Fisher RR, Glatz CE (1988) Biotech Bioeng 32: 777
53. Rothstein F (1978) Technological Problems in Large-Scale Plasma Fractionation. In: Proc. Int'l. Workshop for Protein Separation and Improvement of Blood Plasma Fractionation (Sandberg HE (ed)) USDHEW Publ No. (NIH) 78-1422, p 361, Washington, DC: U.S. Gov't Printing Office
54. Kirwan DJ (1991) Some Problems in the Crystallization of Biological Chemicals. In: Proc. Opportunities and Challenges in Crystallization Research, ISU-ERI-Ames 91150, p 219, Iowa State University
55. Bell DJ, Hoare M, Dunnill P (1983) Adv Biochem Eng Biotech 26: 1 (1983)
56. Bell DJ, Dunnill P (1982) Biotech Bioeng 24: 1271
57. Glatz CE, Hoare M, Landa-Vertiz J (1986) AIChE J 32(7): 1196
58. Larsson P-O, Flygare S, Mosbach K (1984) In: Methods in Enzymology 104: 364
59. Shieh J-Y, Glatz CE (submitted 1991) Precipitation of proteins with polyelectrolytes: Role of polymer molecular weight. In: Dubin P et al. (eds) Polymer soluble complexes Springer, Berlin. Heidelberg New York
60. Glatz CE (in press) In: Solubility of Protein Pharmaceuticals (Ahern TJ, Manning MC (eds)), New York: Plenum
61. Hansen PMT, Hidalgo J, Gould IA (1971) J Dairy Sci 54: 830

62. Melling J, Atkinson A (1972) J Appl Chem Biotechnol 22: 739
63. Atkinson A, Jack GW (1973) Biochem Biophys Acta 308: 41
64. Bonnerjea J, Jackson J, Hoare M, Dunnill P (1988) Enzyme Microb Technol 10: 357
65. Hoare M, Dunnill P (1989) Phil Trans R Soc Lond 324: 497
66. Dunnill P, Lilly MC (1972) Biotech Bioeng Symp 3: 97
67. Agerkvist I, Enfors S-O: Paper presented at 32nd IUPAC Congress, Stockholm 1989
68. Cordes RM, Sims WB, Glatz CE (1990) Biotech Bioeng 6: 283
69. Parker DE, Glatz CE, Ford CF, Gendel SM, Suominen IA, Rougvie MA (1990) Biotech Bioeng 36: 467
70. Belfort G (1989) Ber Bunsenges Phys Chem 93, 939
71. Georgiou G (1988) AIChE J 34: 1233
72. Sharma SK (1986) Separation Sci Tech 21: 701
73. Hammond PM, Atkinson T, Sherwood RF, Scawen MD (1991) BioPharm 4(April): 16
74. Uhlen M, Moks T (1990) Gene Fusions for Purpose of Expression: An Introduction In: Methods in Enzymology, Vol 185, (Goeddel DV (ed)), New York: Academic
75. Enfors S-O, Hellebust H, Köhler K, Strandberg L, Veide A (1990) In: Fiechter A (ed) Advances in Biochemical Engineering/Biotechnology vol 43, p 31, Springer, Berlin Heidelberg New York
76. Ford CF, Suominen IA, Glatz CE: Prot Expr Purif, in press, 1991
77. Ladisch MR, Willson RC, Painton C-D, Builder SE (eds): Protein Purification – From Molecular Mechanisms to Large-Scale Processes, Washington DC: ACS 1990
78. Sherwood R (1990) Trends Biotech 9: 1
79. Itakura K, Hirose T, Crea R, Riggs AD (1977) Science 198: 1056
80. Goeddel DV, Kleid DG, Bolivar F, Heyneker HL, Yansura DG, Crea R, Hirose T, Kaszewski A, Itakura K, Riggs AD (1979) Proc Natl Acad Sci USA 76: 106
81. Shine J, Fettes I, Lan NCY, Roberts JL, Baxter JD (1980) Nature 285: 456
82. Stanley KK, Luzio JP (1984) EMBO J 3: 1429
83. Moks T, Abrahamsén L, Holmgren E, Bilich M, Olsson A, Uhlén M, Pohl G, Sterky C, Hultberg H, Josephson S, Holmgren A, Jörnvall H, Nilsson B (1987) Biochemistry 26: 5239
84. Hammarberg B, Nygren P-A, Holmgren E, Elmblad A, Tally M, Hellman U, Moks T, Uhlén M (1989) Proc Natl Acad Sci USA 86: 4367
85. Sassenfeld HM, Brewer SJ (1984) Bio Technol 2: 76
86. Brewer SJ, Sassenfeld HM (1985) TIBTECH 3: 119
87. Zhao J, Ford CF, Glatz CE, Rougvie MA, Gendel SM (1990) J Biotech 14: 273
88. Niederauer MQ, Glatz CE, Suominen IA, Ford CF, Stachon DS, Rougvie MA: Paper presented at the AIChE Annual Meeting, Los Angeles 1991
89. Flygare S, Griffin T, Larsson P-O, Mosbach K (1983) Anal Biochem 133: 409
90. Larsson PO, Mosbach K (1981) Biochem Soc Trans 9: 285
91. Nguyen AL, Luong JHT (1989) Biotechnol Bioeng 34: 1186
92. Green NM, Konieczny L, Toms EJ, Valentine RC (1971) Biochem J 125: 781
93. Morris JE, Fisher RR: Paper presented at the Annual AIChE Meeting, Paper 78H, 1989
94. Senstad C, Mattiasson B (1989) Biotechnol Bioeng 33: 216
95. Senstad C, Mattiasson B (1990) Biotechnol Bioeng 34: 387
96. Taniguchi M, Kobayashi M, Natsui K, Fujii M (1989) J Ferment Bioeng 68: 32
97. Bradshaw AP, Sturgeon RJ (1990) Biotech Techniq 4: 67
98. Guzman RZ, Kilpatrick PK, Carbonell RG. In: Downstream Processing and Bioseparation: Recovery and Purification of Biological Products, ACS Symposium Series (Hamel J-FP (eds)), vol. 419, p 212, ACS 1990
99. Clonis YD (1988) CRC Crit Rev Biotechnol 7: 263
100. Lowe CR, Pearson JC: In: Affinity Chromatography and Biological Recognition (Chaiken IM, Wilchek M, Parikh I (eds)) London: Academic 1983
101. Hayet M, Vijayalakshmi MA, (1986) J Chromatography 376: 157
102. Morris JE, Fisher RR (1990) Biotech Bioeng 36: 737
103. Pearson JC, Burton SJ, Lowe CR (1986) Analyt Biochem 158: 382
104. Pearson JC, Clonis YD, Lowe CR (1989) J Biotech 11: 267
105. Johansson G, Joelsson M (1986) Anal Biochem 158: 104
106. Smith MC, Cook JA, Furman TC, Gesellchen PD, Smith DP, Hsiung H. In: Protein Purification: From Molecular Mechanisms to Large-Scale Processes (Ladisch MR, Willson RC, Painton CC, Builder SE (eds)), vol 427, p 168, Washington DC: ACS 1990
107. Hochuli E, Bannwarth W, Döbeli H, Gentz R, Stüber D (1988) Bio/Technol 6: 1321
108. Sternberg M (1970) Biotechnol Bioeng 12: 1

109. Jendrisak J. In: Protein Purification: Micro to Macro (Burgess R (ed)), p 75, Alan R Liss, 1987
111. Bozzano AG, Glatz CE (1991) J Membrane Sci 55: 181
112. Naeher G, Thum W, Production of Enzymes for Research and Clinical Use. In: Industrial Aspects of Biochemistry (Spencer B (ed)), p 47, New York: Elsevier 1974
113. Clark KM, Glatz CE (1987) Biotech Prog 4: 241
114. Hidalgo J, Hansen PMT (1969) J Agric Food Chem 17: 1089
115. Li JKK, Johnson T, Yang Y, Shore V (1989) J Virol Meth 26: 3
116. Sternberg M, Chiang JP, Eberts NJ (1976) J Dairy Sci Technol 13: 61
117. Hill RD, Zadow JG (1974) J Dairy Res 41: 373
118. Hill RD, Zadow JG (1978) J Dairy Res 45: 77
119. Anderson AJ (1967) Biochem J 104, 18
120. Shibata T, Cunningham RP, Rudding CM (1984) J Biol Chem 256: 7557
121. Hill RD, Zadow JG (1978) NZ J Polymer Sci Technol 13: 61
122. Gault NFS, Lawrie RA (1980) Meat Sci 4: 167
123. Cotton FA, Wilkinson G (1966) Advanced Inorganic Chemistry, Second Ed, p 941, New York: Interscience
124. Burgess RR, Jendrisak JJ (1975) Biochemistry 14: 4634
125. Jendrisak JJ, Burgess RR (1975) Biochemistry 14: 4639

# Author Index Volumes 1–47

# Subject Index